// The Internet of Us //

Also by Michael Patrick Lynch

In Praise of Reason: Why Rationality Matters for Democracy

Truth as One and Many

True to Life: Why Truth Matters

Truth in Context: An Essay on Pluralism and Objectivity

// The.
Internet
of. Us //

Knowing More and Understanding Less in the Age of Big Data

Michael Patrick Lynch

Liveright Publishing Corporation
A Division of W. W. Norton & Company
Independent Publishers Since 1923
New York   London

For information about permission to reproduce selections from this
book, write to Permissions, Liveright Publishing Corporation
a division of W. W. Norton & Company, Inc., 500 Fifth Avenue,
New York, NY 10110

For information about special discounts for bulk
purchases, please contact W. W. Norton Special Sales at
specialsales@wwnorton.com or 800-233-4830

Manufacturing by RR Donnelley, Harrisonburg, VA
Book design by Daniel Lagin
Production manager: Anna Oler

Library of Congress Cataloging-in-Publication Data
Names: Lynch, Michael P. (Michael Patrick), 1966– author.
Title: The Internet of us : knowing more and understanding less in
the age of big data / Michael Patrick Lynch.
Description: First Edition. | New York : Liveright Publishing
Corporation, 2016. | Includes bibliographical references and index.
Identifiers: LCCN 2015051171 | ISBN 9780871406613 (hardcover)
Subjects: LCSH: Knowledge, Theory of. | Information technology. |
Internet.
Classification: LCC BD161 .L88 2016 | DDC 001—dc23 LC record
available at http://lccn.loc.gov/2015051171

Liveright Publishing Corporation
500 Fifth Avenue, New York, N.Y. 10110
www.wwnorton.com

W. W. Norton & Company Ltd.
Castle House, 75/76 Wells Street, London W1T 3QT

2 3 4 5 6 7 8 9 0

For Rene

In the past, the things that men could do were very limited. . . . But with every increase in knowledge, there has been an increase in what men could achieve. In our scientific world, and presumably still more in the more scientific world of the not distant future, bad men can do more harm, and good men can do more good, than had seemed possible to our ancestors even in their wildest dreams.

—Bertrand Russell

All I know is that I don't know.
All I know is that I don't know nothing.

—Operation Ivy

Contents

Contents

Contents

Preface

The changes wrought by the Internet are sometimes compared to those brought about by the printing press. In both cases, technological advances led to new ways of distributing information. Knowledge became more widely and cheaply available, which in turn led to mass education, new economies and even social revolution.

But in truth, the comparison with the printing press underplays the significance of the changes being brought about by the Internet today. The better comparison is with the written word.

Writing is a technology, a tool. Yet its invention wasn't just a change in how information and knowledge was distributed. It was a new way of knowing itself. Writing allows us to communicate across time—both with ourselves and with others. It allows us to outsource memory tasks and therefore lessen our cognitive load.

Not long ago, for example, I discovered a note my father had written, taped to the back of an old chainsaw I had inherited from him. It was more than a note, really; it was a little essay, detailing good and bad practice with the saw. My dad's house was peppered with such memos. He would write them as a reminder of how best to go about various tasks that one might do only irregularly—replacing the fuel filter on the lawn mower, shutting down the water heater. He would then tape them in a spot where he would be sure to later run across them. When I was a teenager, I found it embarrassing, but I get it now. He was a busy man and knew that he might forget a trick or lesson he'd learned while doing something for the first time. He was, in short, communicating with his future self, while simultaneously relieving his present self of the burden of remembering. That, in microcosm, is what writing allows us to do, and also why its invention is one of the most important developments in human history. It allows us to time-travel and share the thoughts of those who have come before.

The Internet is bringing about a similar revolution in our ways of knowing. Where the written word allows us to time-travel, the Internet allows us to teleport—or at least to communicate in an immediate way across spatial gulfs. Changes in information technology are making space increasingly irrelevant. Our libraries are no longer bounded by physical walls, and our ways of processing and accessing what is in those libraries don't require physical interaction. As a result, we no longer have to travel anywhere to find the information we need. Today, the fastest and easiest way of knowing is Google-knowing, which means not just "knowledge by search engine" but the way we are increasingly dependent on knowing via digital means. That can be a good thing; but it can also weaken and undermine other ways of knowing, ways that require more creative, holistic grasps of how information connects together.

New technology has always spurred a similar debate—and it should. During the heyday of postwar technological expansion in the 1950s, philosophers and artists worried about what that nuclear weapon technology was doing to us, and whether our ethical thinking was keeping up with it. Bertrand Russell, writing in the *Saturday Evening Post*, argued that we need more than expanded access to knowledge; we need wisdom, which he took as a combination of knowledge, will and feeling.[1] Russell's point was simple: growth in knowledge without a corresponding growth in wisdom is dangerous. This book is motivated by a similar worry and with a desire to do something about it. Yet where Russell was concerned with a specific kind of knowledge —knowledge of nuclear bombs—my concern is with the expansion of knowledge itself, with how the rapid changes in technology are affecting how we know and the responsibilities we have toward that knowledge.

Still, this is not an "anti-technology" book. I'm a dedicated user of social media and the platforms that enable it (the rise of which is sometimes called "Web 2.0"). I tweet, I Facebook, I have a smartphone, a tablet, and more computers than I care to admit. I am in no position to write an anti-technology book. Technology itself is not the problem. Unlike nuclear weapons or guns, information technology itself is generally not designed to kill people (although it can certainly lend a hand). Information technologies are more like cars: *so fast, sleek and super-useful*

Preface

that we can overrely on them, overvalue them and forget that
their use has serious consequences. The problems, such as they
are, are due to how we are using such technologies.

My aim is to examine the philosophical foundations of what
I'll call our digital form of life. And whether or not my conclusions are correct, it is clear that this is a task we must engage in
if we want to avoid the fate that worried Russell: being swallowed
up by our technology.

Storrs, CT
October 2015

// Part I.

The *New* Old Problems of Knowledge

1.

Our Digital Form of Life

Neuromedia

Imagine a society where smartphones are miniaturized and hooked directly into a person's brain. With a single mental command, those who have this technology—let's call it neuromedia— can access information on any subject. Want to know the capital of Bulgaria or the average flight velocity of a swallow? It's right there. Users of neuromedia can take pictures with a literal blink of the eye, do complex calculations instantly, and access, by thought alone, the contact information for anyone they've ever met. If you are part of this society, there is no need to remember the name of the person you were introduced to last night at the dinner party; a sub-cellular computing device does it for you.

For the people of this society, it is as if the world is in their heads. It is a connected world, one where knowledge can be instantly shared with everyone in an extremely intimate way. From the inside, accessing the collective wisdom of the ages is as simple as accessing one's own memory. Knowledge is not only easy; everyone knows so much more.

Of course, as some fusspots might point out, not all the information neuromedia allows its users to mentally access is really "knowledge." Moreover, they might claim, technological windows are two-way. A device that gives you a world of information also gives the world huge amounts of information about you, and that might seem like a threat to privacy. Others might fret about fragmentation—that neuromedia encourages people to share more information with those who already share their worldview, but less with those who don't. They would worry that this would make us less autonomous, more dependent on our particular hive-mind—less human.

But we can imagine that many in the society see these potential drawbacks as a price worth paying for immediate and unlimited access to so much information. New kinds of art and experiences are available, and people can communicate and share their selves in ways never before possible. The users of neuromedia are not only free from the burden of memorization, they are free from having to fumble with their smartphone, since thoughts can be uploaded to the cloud or shared at will. With neuromedia, you have the answer to almost any question immediately without effort—and even if your answers aren't always right, they are right most of the time. Activities that require successful coordination between many people—bridge building, medicine, scientific inquiry, wars—are all made easier by such pooled shared "knowledge." You can download your full medical history to a doctor in an emergency room by allowing her access to your own internal files. And of course, some people will become immensely wealthy providing and upgrading the neural transplants that make neuromedia possible. All in all, we can imagine, many people see neuromedia as a net gain.

Now imagine that an environmental disaster strikes our invented society after several generations have enjoyed the fruits of neuromedia. The electronic communication grid that allows neuromedia to function is destroyed. Suddenly no one can access the shared cloud of information by thought alone. Perhaps backup systems preserved the information and knowledge that people had accumulated, and they can still access that information in other ways: personal computers, even books can be dusted off. But for the inhabitants of the society, losing neuromedia is an immensely unsettling experience; it's like a normally sighted person going blind. They have lost a way of accessing information on which they've come to rely. And that, while terrible, also reveals a certain truth. Just as overreliance on one sense can weaken the others, so overdependence on neuromedia might atrophy the ability to access information in other ways, ways that are less easy and require more creative effort.

While neuromedia is currently still in the realm of science fiction, it may not be as far off as you think.[1] The migration of technology into our bodies—the cyborging of the human—is no longer just fantasy.[2] And it shouldn't surprise anyone that the possibilities are not lost on companies such as Google: "When you think about something and don't really know much about it, you will automatically get information," Google CEO Larry Page is quoted as saying in Steven Levy's recent book *In the Plex*.

"Eventually you'll have an implant, where if you think about a fact, it will just tell you the answer."[3]

This possibility raises some disquieting questions about society, identity and the mind. But as Larry Page's remark suggests, the deeper question is about information and knowledge itself. How is information technology affecting what we know and how we know it? And what happens to society if we not only know more about the world but the world knows more about us? Taken seriously, these questions force us to grapple not only with *how* we know with technology, but with how we *should*. That's the really important problem, and it is the philosophical and ethical question at the core of this book—one I'll argue we ignore at our peril.

My hypothesis is that information technology, while expanding our ability to know in one way, is actually impeding our ability to know in other, more complex ways; ways that require 1) taking responsibility for our own beliefs and 2) working creatively to grasp and reason how information fits together. Put differently, information technologies, for all their amazing uses, are obscuring a simple yet crucial fact: greater knowledge doesn't always bring with it greater understanding.

So, a large part of my aim in this book is to explore how the Internet is changing our minds and lives. That it is doing so is beyond doubt. If you are like me, you already feel a lot smarter when you have access to Google, and somewhat frustrated when you do not—in almost the exact way you feel when you suddenly can't remember something you knew just yesterday. Knowing by Google is now so familiar that it has an unnoticed seamlessness that we earlier only attached to perception. Where we used to say that seeing is believing, now we might say "googling is believing." And yet this very fact also makes it easier for people to believe that Barack Obama is a Muslim, or that the measles vaccine is dangerous. Just as we often see what we want to see, we often google what we want to google.

The increasingly seamless integration of our digital experiences into our lives is not the result of a single shift but the result of a gradual series of changes. Internet wonks tend to think that we are seeing the arrival of the "third wave" of the Internet. First there was Web 1.0 (the ancient days of "Wow! You should check out this email thing!"). Then, starting in the early 2000s, there was Web 2.0. ("Wow! You should check out this Facebook thing!"). Now we have Web 3.0 (the "smart Web") and, most significantly, the so-called Internet of Things ("Wow! You should check out my smart . . . watch, refrigerator, lamp, socks!").

In essence, the "Internet of Things" is a way of describing the phenomenon of networked objects—objects that are embedded with data-streaming sensors and software that connect them to the Net. The "things" in question run the gamut from autonomous connected devices like smartphones to the tiny radio-frequency identification (RFID) microchips and other sorts of sensors attached to everything from UPS trucks and cargo containers to pets, farm animals, cars, thermostats, and NFL helmets. By 2007 there were already 10 million sensors of all sorts connected to the Internet, and some projections have that number rising to 100 trillion by 2030 if not before.[4] These sensors are being used not only for economic purposes but for scientific ones (to track migratory animals, for example), and for security and military purposes (such as tracking human beings). According to Jeremy Rifkin, a leading economist of the digital world, the Internet of Things is even giving rise to a "Third Industrial Revolution," precipitating huge changes in how human beings around the globe interact with one another, economically and otherwise.[5]

The Internet of Things is made possible by—and is also producing—big data. The term "big data" has no fixed definition, but rather three connected uses. First, it names the ever-expanding volume of data that surrounds us. You've heard some of the statistics. As long ago as 2009, there were already 260 million page views per month on Facebook; in 2012, there were 2.7 billion likes per day. An estimated 130 million blogs exist; there are around 500 million tweets per day; and billions of video views on YouTube. By some estimates, the amount of data in the world in 2013 was already something around 1,200 exabytes; now it is in the zetabytes. That's hard to get your mind around. As Viktor Mayer-Schönberger and Kenneth Cukier estimate in their recent book, *Big Data: A Revolution That Will Transform How We Live, Work, and Think*, if you placed that much information on CD-ROMs (remember them?) it would stretch to the moon five times. It would be like giving every single person on the earth 320 times as much information as was stored in the ancient library of Alexandria.[6] And by the time you are reading this, the numbers will be even bigger.

So, one use of the term "big data" refers to the massive amount of data making up our digital form of life. In a second sense, it can be used to talk about the analytic techniques used to extract useful information from that data. Over the last several decades, our analytic methods for information extraction have increased in sophistication along with the increasing size of the data sets we have to work with. And these techniques have been put to a mind-boggling assortment of uses, from Wall Street to science of all sorts. A simple example is the data "exhaust" you are leaving as you read these very words on your Kindle or iPad. How much of this book you read, the digital notes you take on it, is commercially available information, extracted from the trail of data you leave behind as you access it in the cloud. Booksellers like Barnes and Noble and Amazon can, and have, used this sort of information to further target the types of products they market.

As a consequence of the increasing importance of data analytics, we might employ "big data" in a third sense—to refer to firms like Google or Amazon that utilize data analytics as an essential part of their business model, and government agencies like the NSA that use these techniques as an essential part of, well, *their* business model. In this third sense, Big Data is like Big Oil. Large oil conglomerates are powerful because they control how the world's major energy resource is not only distributed but how it is extracted. The tech giants are similar. Energy is not information, but both are resources, and resources by which the world runs. And Big Data, like Big Oil, is big precisely because it can control access to data as well as the extraction of information and knowledge from that data. Big Data refines data for information and knowledge, and we need to pay attention to that fact because knowledge, like energy, is not just a passive, inert resource. It is fuel: fuel for our ideas, our actions, everything. And the power that comes with control over that fuel is therefore formidable. Knowledge, as Sir Francis Bacon said, is power.

The big numbers behind big data, and the power inherent in those numbers, are impressive. Not long ago, it was said we were living in a time of information "glut"; we were "flooded"; we were "overloaded." While some still feel this way, for most of us, the sense of being overwhelmed by information is passing. Digital data is something that is no longer drowning us. We are adapting to life under water, we are breathing it all in, becoming digitally human. Information is the atmosphere—what the philosopher Luciano Floridi calls the infosphere—of our lives.[7] But the fact that we live in the infosphere, that it is becoming ordinary, doesn't mean that we understand it, nor how it is changing us and what Ludwig Wittgenstein might have called our form of life. A form of life, as I mean it here, is the myriad practices of a culture that create their philosophies, but also, in Stanley Cavell's words, their "routes of interest and feeling, sense of humor, and of significance, and of fulfillment of what is outrageous, of what is similar to what else."[8] As I read him, Wittgenstein thought that once a set of practices is ingrained enough to become your form of life, it is difficult to substantively critique them or even to recognize them as what they are. That's because our form of life is "what has to be accepted, the given."[9] We can no longer get outside of it.

One way of describing the direction in which our own culture is moving is that many of us are starting to adapt what we might call a digital form of life—one which takes life in the infosphere for granted, precisely because the digital is so seamlessly integrated into our lives. The Internet of Things is becoming the Internet of Us, and figuratively, if not yet literally, we are becoming digital humans.

What is amazing is not *that* this is happening, but *how quickly* it is happening, how quickly we are settling in and accepting our new ways of being. That is particularly true, I think, with regard to our new practices of knowing. If anything, recent years have seen a rushing tide of enthusiasm. We've been told that the Internet has been a force for undiluted knowledge expansion and democratization. Not only do we know more, but more people know. Our minds work faster, multitask more, and just plain get more stuff done.[10]

William James famously said that once a current of thought like this starts to surge, there is little you can do. Trying to stop it is like planting a stick in a river, "round your obstacle flows the water and 'gets there just the same.'"[11] He's got a point, but I endeavor to plant another stick in the river anyway—not because I am unhappy with my iPhone, or hostile to the growth of knowledge, but for a simpler reason. Acceptance without reflection is dangerous, and while our stick may not stop the flow, it can help us measure and assess its depth and direction. As the literary critic and writer Leon Wieseltier remarks, "every technology is used before it is completely understood."[12] There is a lag time, and when we are living in the midst of a lag, that is precisely when we need to pay attention: before the river becomes settled in its course, something we take for granted as part of the natural landscape.

For another example, think about cars. The automobile remains an incredible invention. It increased autonomy, allowed for the distribution of goods and services into remote areas and driving one can be a lot of fun. But needless to say, our unthinking commitment to the technology—our willingness from early on to let it swamp other technologies, to treat it as having more value than other means of transportation—has had seriously negative consequences as well. The devaluing (at least in the United States) of public transportation systems like trains and the rise of carbon emissions and pollution are just the more obvious examples. In the United States especially, it has been difficult for us, as a culture, to come to grips with these problems. The technology has become so embedded, so part of our form of life, that we have a hard time even noticing how dependent on it we really are.

In the same way, paying attention to our digital form of life—seeing it for what it is, both good and bad—is easier said than done. Forms of life are complicated and filled with contradictions. That's true of our emerging digital form of life too. We digital humans do have access to more information than ever before—whether or not we have neuromedia. But it is also true that in other respects we know less, that the walls of our digital life make real objective knowledge harder to come by, and that the Internet has promoted a more passive, more deferential way of knowing.[13] Like our imaginary neuromedians, we are in danger of depending too much on one way of accessing the world and letting our other senses dull.

Socrates on the Way to Larissa

Data is not the same thing as information. As the founding father of information theory, Claude Shannon, put it back in the 1940s, data signals are noisy, and if you want to filter out what's meaningful from those signals, you have to filter out at least some of that noise. For Shannon the noise was literal. His groundbreaking work concerned how to extract discernible information from the signals sent across telephone lines.[14] But the moral is entirely general: bits of code aren't themselves information; information is what we extract from those bits. They are the meaningful leftovers after we filter out the noise.

Yet not all information is good information; information alone still doesn't amount to knowledge. So, what is knowledge?

If you want to find out what anything really is—what knowledge is, in this case—a really good way to begin is to ask why anyone should give a damn about it. Plato himself asked this question in a famous dialogue from the third century BCE, where he imagines his teacher Socrates asking about why it matters that someone should know, rather than merely guess, directions to Larissa. Today as then, Larissa is a busy cultural and urban center, nestled in the mountainous Greek region of Thessaly. Legend had it that Achilles founded the city, and Hippocrates, the famous physician, supposedly died there. It was also the birthplace of the Greek general Meno—a man now more famous for having the starring role in this particular dialogue than any military victory.

Near the end of Plato's piece, Socrates asks Meno: Why does knowledge matter anyway? His questioning is pointed. In particular, he wants Meno to tell him why knowledge matters more than "true opinion." After all, Socrates says, if I ask some passing stranger directions to Larissa, we'll get there as long as he has a true opinion about the matter—even if it is a lucky guess. I won't get there any faster by asking someone who really "knows" the answer—such as someone who has traveled there before. And that brings us to Socrates' inquiry: why does knowledge seem to matter so much since having accurate information can often get us to where we want to go? Meno fumbles about, and uncharacteristically, Socrates himself is quick to offer an answer in the form of a metaphor. Opinions without knowledge—even true ones—he says, are like the statues of Daedalus: so lifelike that they would get up and walk away if not tied to the ground. Knowledge, he seems to suggest, is true opinion that is tied down or grounded.

Plato's dialogue illustrates three simple points that are good to keep in mind when thinking about knowledge (what the

Greeks sometimes called *epistêmê*, from which we get the word epistemology, or the study of knowledge). It is worth getting these points out in front.

Knowing something is different from just having an opinion about it. Any old fool can opine, but few can know. We might put this another way by saying that mere information or data isn't knowledge; information can be better or worse, accurate or inaccurate. When we want to know, we want the right or true information. But we also want something more.

Having accurate information still isn't enough to know either. Making a lucky guess isn't the same as knowing. The lucky guesser doesn't have any ground or justification for his opinion, and as a result, he is not a *reliable* source of information on that topic. Ask him again tomorrow and he might guess something else. That's why his information is ultimately less valuable in most situations. When we want to know we want more than guesses; we want some sort of basis for trust.

What grounds our opinions or beliefs matters for action. The old intelligence services adage is that knowledge is actionable information. Actionable information is information you can work with—that, in short, you can trust. Guesses are not actionable—even if they are lucky, precisely because they are *guesses*. What's actionable is what is justified, what has some ground.

So: whatever else it is, knowing is having a correct belief (getting it right, having a true opinion) that is grounded *or justified*, and which can therefore guide our action. Call this *the minimal definition of knowledge.*

The minimal definition of knowledge is helpful to a point. But like a lot of pithy definitions, it obscures as well as illuminates. In particular, it passes over the fact that how a belief is grounded comes in different forms. Suppose I ask you the best way to get to Larissa and you give me the correct answer, not because you guess but because you have some grounds for it. There are lots of different ways that could happen. For example, you might:

- Look at the map on your phone.
- Recall how you got there last year.
- Do both of these things but also explain why certain routes that look good on the map are actually slower because of localized road construction, etc.

All three of these points might allow you to know, but in different ways. They represent three different ways our opinions can be grounded, by being based on:

- Reliable sources.
- Experience or reasons that we possess.
- A grasp of the big picture.

The first sort of knowing is the sort we do when we absorb information from expert textbooks or good Internet resources. The second is the sort of knowing we value whenever possessing reasons or experience matters. And the third is different still—it is the sort of knowing we expect of our most creative experts— even if those experts are more intuitive than discursive in their abilities. This is what I'll call understanding.

Understanding, as in our example, often incorporates the other ways of knowing, but goes farther. It is what people do when they are not only responsive to the evidence, they have creative insight into how that evidence hangs together, into the explanation of the facts, not just the facts themselves. Understanding is what we have when we know not only the "what" but the "why."[15] Understanding is what the scientist is after when trying to find out *why* Ebola outbreaks happen (not just predict how the disease spreads). It is what you are after when trying to understand *why* your friend is so often depressed (as opposed to knowing that she is).

In real life, all the ways we have of knowing are important. But without understanding, something deeper is missing. And our digital form of life, while giving us more facts, is not particularly good at giving us more understanding. Most of us sense this. That is one reason we try to limit our children's screen time and encourage them to play outside. Interaction with the world brings with it an understanding of how and why things happen physically that no online experience can give. And it is why so many of us who use Facebook are still troubled by its siren song: it is a simulacrum of intimacy, a simulacrum of mutual understanding, not the real thing. The pattern of what people like or don't like tells us something about them—more, in fact, than they may wish. But it doesn't tell us why they like what they like. It doesn't allow us to understand them. Facebook knows, but doesn't understand.

As we'll see in more detail later, understanding not only gets us the "why," it brings with it the "which"—as in which question to ask. Those who know, do. But those who understand also ask the right question—and therefore can find out what to do next. Asking questions was Socrates' special skill. It is perhaps for that reason that the Oracle of Delphi famously told Socrates that he was the wisest man in Athens. According to Plato, Socrates himself said that all he knew was that he didn't know much. And maybe he didn't. But one can't read Plato without thinking that the Oracle was on to something. Socrates was a champion not of knowledge per se, but of understanding. That's the skill we need to remember now. It may sound trite but it is true nonetheless: we need to rediscover our inner Socrates.

Welcome to the Library

In one of his most famous stories, the great Argentinian writer Jorge Luis Borges imagined what it would be like to live in a world comprised of a single, almost infinite library, containing a virtually uncountable number of books, ranging from tomes of incomprehensible nonsense to treatises on everything from politics to particle physics. In one way, the library seems to make knowledge easy; all the truths of the world would be at your disposal. Of course, so would many falsehoods. And if you lived there, the library would be all you knew. There would be no escaping its walls to find an independent check—no way, except by appeal to the library itself, of knowing which books contain the truth and which do not.

The story I want to tell is the story of how our culture is dealing with the fact that most of us are living in the library now—the Library of Babel, as Borges dubbed it in the short story by that name. It is a virtual library, and one that may indeed migrate right into our brain, should neuromedia ever come to pass. But whether or not that happens, the story I have to tell is an unapologetically philosophical one. As Borges knew more than most, our philosophies are part of what make up our culture, our form of life, and we need to come to grips with them if we want to understand ourselves.

Even in a story of ideas, central characters have a backstory. The backstory of our ideas about knowledge is that they've grown up shaped by some very ancient problems, problems that the surging changes in information technology are dragging to the surface of our cultural consciousness and casting in new forms.

Compare, for example, neuromedia with an old philosophical chestnut—the thought experiment of the Brain in the Vat. It goes like this: How do you know that you aren't simply a brain hooked up to a computer that is busily making it seem as if you have a body, and are reading this book (and thinking about brains in vats)? If so, the world is just an illusion, manufactured for your benefit, and almost everything you think you know is actually false. This is the philosophical position known as skepticism. The skeptic's basic idea is that we can't ever determine whether what seems to us to be the case really is true. If so, then either we punt on knowing what is true, or punt on truth itself.

The Brain in the Vat is itself an updated version of Descartes' seventeenth-century story of the evil demon who spends all his time deceiving us. The idea is that all of our experience might be misleading. Descartes thought that even the demon (or the lab-coated evil scientist running the vat) couldn't fool us about everything: for he can't trick you into thinking you don't exist without your existing already. Nonetheless, it seems like he could trick you about almost everything else. If the illusion is so perfect, what further experience could prove to us that our experiences are illusory? Whatever experience it is (including Laurence Fishburne turning up in cool shades with blue and red pills in

The Matrix) would just be *more experience*. It could be an illusion too. And that will hold no matter how hard we work at gathering data, how open-minded we are to new information, or how objective we are in considering the facts. If the world is but a dream, then so too is our best science. And if that is possible, then maybe we don't really know most of what we think we do.

Now think about our two stories—neuromedia and the Brain in the Vat—together for a minute. In some ways they are mirror images. The one puts the computer in your brain; the other puts your brain in the computer. The one appears to make knowledge easy, the other makes it impossible. But look closer and you'll see that they are more alike than they appear at first glance. They raise some of the same underlying philosophical questions. For example, in our earlier discussion, we were *assuming* that much of the information you'd be able to access through neuromedia is true. But how would we know that exactly? By checking neuromedia? Of course, we might ask someone else. But if everyone—or at least the people nearby—are also hooked up to the same sources, then it is not clear what we would really know. *In both cases, it seems, real knowledge—knowledge of what is the case as opposed to what we just happen to think is the case—is possible only by escaping the machine and getting to the world "outside."*

Yet what if there is no getting "outside" the machine? What if even "brains in vats" aren't real, and we are all just living a *completely* simulated life? That worry is closer to Descartes' original nightmare. Its closest contemporary analogue is the thought that you and I are really just SIMs. A SIM is a "simulated person"—simulated by a computer program, for example. SIMs already exist. Popular Web-based games like Second Life, for example, have allowed people to create artificial "people" with SIM backgrounds, jobs, spouses, etc., for years. These programs even allow your SIM to continue to interact with other SIMs when you aren't actively playing the game, pursuing its career, relationships and so on. And that, as the philosopher Nick Bostrom has recently suggested, raises the possibility that the universe in which we live is and always has been a simulation run by a computer program created for the amusement of super-beings with superior technology.[16] If so, then we aren't just wrong about whether, for example, we have arms and legs (as opposed to just being brains in vats). We are wrong about the nature of our universe itself: we might be living in a universe completely composed of information, whose underlying particles are really just the 1s and 0s of computer code.

Thought experiments like this show that ancient skeptical worries about knowledge are not only still with us, they are being made anew. But really, we don't need to appeal to SIMs, Brains in Vats or nearly infinite libraries to see that our culture is facing an intellectual crisis about knowledge. As we'll see over the next few chapters, the new "old" philosophical problems that make up the backstory of our form of life are actually far more immediate, more pressing, and less abstract. That's what makes them so unnerving.

2.

Google-Knowing

Easy Answers

One day in the summer of 2014, I wrote down four questions to which I didn't know (or had forgotten) the answers. The challenge: to answer the questions without relying—at all—on the Internet. The four questions were:

1. What is the capital of Bulgaria?
2. Is a four-stroke outboard engine more efficient than a two-stroke?
3. What is the phone number of my U.S. representative?
4. What is the best-reviewed restaurant in Austin, Texas, this week?

Number 1, unsurprisingly, was the easiest. I suspected it was Sofia, and a map of Europe and a large reference dictionary I had in the house confirmed it. (I was briefly worried about how up-to-date the information was, as the dictionary was almost two decades old, the map older.)

Question 2 proved more difficult. I had a (nonfunctioning) four-stroke engine, and it had a manual, but it said nothing about the newer two-stroke engines. Some boating reference books I had lying around were of no help. So I went to the local marina and spoke to a mechanic I knew. He was full of information, and had time to give me the basics. I even got to look at an engine. That was great, until I got home and realized I had not taken any notes. I was coming to think that I would make a very poor investigative journalist.

Initially, I had thought number 3 would be the easiest, until I remembered we no longer had a phone book (with the blue government pages). I started to call information, but then wondered whether they'd be using the Internet. Assuming the answer was yes, I stopped in at the local library. The kid behind the counter looked at me funny when I asked. He suggested, more than a little wryly, that I use one of their computers. I countered by asking whether they had any local phone books. They actually did—it was several years old, but still relevant. Mission accomplished.

It was question 4 that stumped me. I knew no one in Austin well enough to call for an opinion. I thought of calling their local chamber of commerce, but I didn't have a way to get that number. Besides, how would they even know the answer to such a question? My library in Connecticut didn't have copies of any Texas papers. Books might help, but the ones I looked at, such as a few travel guides at the local bookstore, were not current enough. In short, I was out of luck.

None of this, I'm sure, surprises you. It is common knowledge that our ways of knowing about the world have changed. Most knowing now is Google-knowing—knowledge acquired online. But my little exercise brought it home for me, made it personal, in a way that I hadn't before appreciated, and I encourage you to try it yourself. It feels historic, something akin to what I imagine it must be like to dress up in period costume and live in a tent, as some Civil War reenactors do.

Just a dozen or so years ago the processes I went through to answer my questions wouldn't have seemed at all out of the ordinary. Research involved footwork, and many academics still doubted the veracity of information acquired online. But that battle is long lost. The Internet is the fountain of knowledge and Google is the mouth from which it flows. With the Internet, my challenge is no challenge at all; answering the questions is easy. Just ask the knowledge machine.

Speed is the most obvious distinguishing characteristic of how we know now. Google-knowing is fast. Yet as my exercise brings home, this speed is so dramatic that it does more than just save time. The engine diagram I can call up on my phone can be consulted again and again. I don't need notes—or need them less, and I can store them on the cloud in case I do. Elected representatives are easier to track down than ever before; I can send my opinion to them (or at least to their addresses and offices) any number of ways and in seconds. Thanks to Google Street View, I can see what Sofia and its inhabitants look like up close and personal. And question 4—a question of a sort that probably wouldn't even have been posed before the Internet—is addressed by any number of sites giving me rankings and reviews of restaurants.

Not everything about Google-knowing is new, however. And that itself is important to appreciate. One humorous illustration of this came in 2013, when the website College Humor asked: what if Google was a guy? The ensuing video was hilarious and a bit disturbing. The questions we ask our search engines ("Hedgehog, cute," "Bitcoin unbuy fast," "college girls?") seem all the more ridiculous (and creepy) once we imagine asking them of an actual person—like an amiable but overworked bureaucrat behind a desk. But it also reminds us of a fact about how we treat Google and other search engines—a fact that is obvious enough but often overlooked. We treat them like personal reference librarians; we ask them questions, and they deliver up sources that claim to have the answers. And that means that we already treat their deliverances as akin—at least at the level of trust—to the deliverances of actual people. Of course, that is precisely why the bit is funny: Google *isn't* a guy (or anyone, male or female); it doesn't create information, it distributes it. Yet this is also why it makes sense for us to treat Google like a person—why the video rings true. The information we get from the links we access via Google is (mostly) from other people. When we trust it, we almost always trust someone else's say-so—his or her "testimony." Indeed, the entire Internet, including, of course, Wikipedia, Facebook, the blogosphere, Reddit, and most especially the Twitterverse, etc., can be described as one giant knowledge-through-testimony machine.

"Do you really know what you're doing, or do you Google-search know?"

Fig. 1. Courtesy of Barbara Smaller/The New Yorker Collection/The Cartoon Bank

So, "Google-knowing" helps describe how we acquire information and knowledge via the testimony machine of the Internet. It is easy, fast *and yet dependent on others.* That is a combination that, at least in this extreme form, has never been seen before. Moreover, and as my exercise from last summer indicates, we can essentially no longer operate without it. I Google-know every day, and I'm sure you do too. But partly as a result, Google-knowing is increasingly swamping other ways of knowing. We treat it as more valuable, more natural, than other kinds of knowledge. That's important, because as we'll see, the human mind has evolved to be receptive to information in certain environments. As a result, we tend to trust our receptive abilities automatically. That makes sense in all sorts of cases, especially when we are talking about the senses—seeing, hearing etc. The problem is that Google-knowing really shouldn't be like that; as the *New Yorker* cartoon implies, we shouldn't trust it as a matter of course.

Being Receptive: Downloading Facts

You want to sort the good apples from the bad. Someone gives you a device and tells you to use it to do the sorting. If the device is reliable, then most of the apples you sort into the good pile will, in fact, be good. And this will be the case whether you possess any recognizable evidence to think it is so or not. As long as the device really does its job, it will give you useful information about apples whether or not you have any idea about its track record, or about how it is made, or even if you can't tell a good apple from a pear, or a hole in the ground.

We all need good apple-sorting devices, and not just to sort apples. If you want to find food and avoid predators, which every organism does, you need a way to sort the good (true) information from the bad (false)—and to do so quickly, mechanically and reliably. Call this *being receptive*. When we know in this way, we are reliably tracking the good apples.

Being receptive is a matter of "taking in" the facts. We are being receptive when we open our eyes in the morning and see the alarm clock, when we smell the coffee, when we remember we are late. As we move about, we "download" a tremendous amount of raw data—data that is processed into information by our sensory and neural systems. This information represents the world around us. And if our visual system, for example, is working as it should, and our representation of the world is accurate—if we see things as they are—then we come to know.

Receptive knowledge isn't "intellectual." It is how dogs, dolphins and babies know. To have this sort of knowledge, you don't have to know that you know, or even be able to spell the word "knowledge" (or know that it is a word)—although if you do, that's okay too. *Receptive states of mind* aim to track the organism's environment, and they are causally connected to the organism's stimuli and behavior. In human animals, we might call these states beliefs, and say that human beliefs can be true or false.

So, knowing by being receptive is something we have in common with other animals, and it is clear we need such an idea to explain how animals (including us) get around in the world. When we explain, for example, why a particular species can protect their nests by leading predators away, we assume they can reliably spot predators.[1] We take them to have the capacity to accurately recognize features of their environment ("predator!") in a non-accidental way. So the following seems like a reasonable hypothesis: having representational mechanisms that stably track the environment is more adaptive than having mechanisms that only work on Tuesday and Thursdays.

This kind of explanation is what biologists call a "just-so" story. It assumes that behavior that contributes to fitness makes informational demands on a species, and that species' representational capacities were, at least in most cases, selected to play that role.[2] But this story's assumptions are widely held. It leaves us with a pretty clear picture: for purposes of describing animal and human cognition, we need to think of organisms as having the capacity to know about the world by being *reliably receptive* to their environment, to act as reliable downloaders.[3]

Here's the crucial point for our purposes: an organism's default attitude toward its receptive capacities—like vision or memory—is *trust*. And that makes sense. Even though we know, for example, that our eyesight and hearing can and do mislead us, perception is simply indispensable for getting around in the world. We can't survive without it. Receptive thought is also non-reflective. We don't think about it. That is because ordinarily, most of our receptive processes tick along under the surface of conscious attention. They don't require active effort. This is most obvious in the case of vision or hearing: as you drive down the road to work, along a route you've traveled many times, you are absorbing information about the environment and putting it into immediate action. As we say, much of this happens on auto-pilot. The processes involved are reliable in most ordinary circumstances, which is why most of us can do something dangerous like drive a car without a major mishap. Of course, each of these processes is itself composed of highly complex sub-processes, and each of those is composed of still more moving parts, most of which do their jobs without our conscious effort. In normal operations, for example, our brain weeds out what isn't coherent with our prior experiences, feelings and what else we think we know. This happens on various levels. At the most basic one, we—again, unconsciously—tend to compare the delivery of our senses, and we reject the information we are receiving if it doesn't match.

This sort of automatic filtering that accompanies our receptive states of mind is described by Daniel Kahneman and other researchers as the product of "system 1" cognitive processes. System 1 information processing is automatic and unconscious, without reflective awareness. It includes not only quick situational assessment but also automatic monitoring and inference. Among the jobs of system 1 are "distinguishing the surprising from the normal," making quick inferences from limited data and integrating current experience into a coherent (or what seems like a coherent) story.[4] In many everyday circumstances, this sort of unconscious filtering—coherence and incoherence detection—is an important factor in determining whether our belief-forming practices are reliable. Think again about driving your car to work. Part of what allows you to navigate the various obstacles is not only that your sensory processes are operating effectively to track the facts, but that your coherence filters are working to weed out what is irrelevant and make quick sense of what is.

Yet the very same "fast thinking" processes that help us navigate our environment also lead us into making predictable and systematic errors. System 1, so to speak, "looks" for coherence in the world, looks for it to make sense, even when it has very limited information. That's why people are so quick to jump to conclusions. Consider: *How many animals of each kind did Moses take into the ark?* Ask someone this question out of the blue (it is often called the "Moses Illusion") and most won't be able to spot what is wrong with it—namely, that it was Noah, not Moses, who supposedly built the ark. Our fast system 1 thinking expects something biblical given the context, and "Moses" fits that expectation: it coheres with our expectations *well enough* for it to slip by. [5] Something similar can happen even on a basic perceptual level; we can fail to perceive what is really there because we selectively attend to some features of our environment and not others. In a famous experiment, researchers Christopher Chabris and Daniel Simons asked people to watch a short video of six people passing a basketball around.[6] Subjects were asked to count how many passes the people made. During the video, a person in a gorilla suit walks into the middle of the screen, beats its chest, and then leaves—something you'd think people would notice. But in fact, half of the people asked to count the passes missed the gorilla entirely.

So, the "fast" receptive processes we treat with default trust *are* reliable in certain circumstances, but they are definitely *not* reliable in all. This is a lesson we need to remember about the cognitive processing we use as we surf the Internet. Our ways of receiving information online—Google-knowing—are already taking on the hallmarks of receptivity. We are already treating it more like perception. Three simple facts suggest this. First, as the last section illustrated, Google-knowing is quickly starting to feel indispensable. It is our go-to way of forming beliefs about the world. Second, most Google-knowing is already fast. By that, I don't just mean that our searches are fast—although that is true; if you have a reasonable connection, searches on major engines like Bing and Google deliver results in less than a second. What I mean is that when you look up something on your phone, the information you get isn't the result of much effort on your part. You are engaging quick, relatively nonreflective cognitive processes. In other words, when we access information online, when we try to "Google-know," we engage in an activity that is composed of a host of smaller cognitive processes ticking along beneath the surface of attention. Third, and as a result of the first two points, we often adopt an attitude of default trust toward digitally acquired information. It therefore tends to swamp other ways of knowing; we pay attention to it more.

That is not surprising. Google-knowing is often (although not always) fast and easy. If you consult a roughly reliable source (like Wikipedia) and engage cognitive processes that are generally reliable in that specific context, then you are being receptive to the facts out there in the world. You are tracking what is true—and that is what being a receptive knower is all about. You may not be able to explain why that particular bit of information is true; you may not have made a study of whether the source is really reliable; but you are learning. So, can't we still say that you are knowing in one important sense?

We can. And we do. But Google-knowing is *knowing* only if you consult a reliable source and your unconscious brain is working the way you'd consciously like it to.

If. There, as always, is the rub.

John Locke Agrees with Mom

The day following the bombing of the Boston Marathon in April 2013, social media was clogged with posts of a man in a red shirt holding a wounded woman. The picture was tragic, and the posts made it more so: they told us that the man had planned to propose to the woman when she finished the marathon—until the bomb went off. Hundreds of thousands of people reposted and tweeted the story, often contributing moving comments of their own.

The story, however, proved to be false. The man had not been planning to propose to the woman. They weren't even acquainted. Nor was it true, as was widely reported even in the "mainstream" media (I heard it on my local NPR station the day of the bombing), that the authorities had purposefully shut down cell phone service in Boston (the system simply was flooded with too much traffic). These were rumors, circulating at the speed of tweet.

Rumors like this are also examples of the widely discussed phenomenon of *information cascades*—a phenomenon to which the Internet and social media are particularly susceptible.[7] Information cascades happen when people post or otherwise voice their opinions in a sequence. If the first expressions of opinion form a pattern, then this fact alone can begin to outweigh or alter later opinions. People later in the sequence tend to follow the crowd more than their own private evidence. The mere fact that so many people prior have voiced a particular opinion— especially if they are in some sense within your social circle—the more likely it is that you'll go with that opinion too, or at least give it more weight. Social scientists (and advertising executives) who have studied this phenomenon have used it to explain not only how information often moves around the Internet, but how and why songs and YouTube videos become popular. The more people have "liked" a video, the greater the chance even more people will like it, and pretty soon you end up with "Gangnam Style" and "What Does the Fox Say?"

Information cascades are hardly new. The mob mentality has worked its dark magic as long as there have been mobs. That's why my mom used to ask, in response to my whine that "everyone else is [doing, saying, believing] something" that, "If everyone jumped off a bridge, would you jump off too?" Well, hopefully not, but the history of humanity might suggest otherwise. We not only tend to follow others' actions, we also seem all too willing to go along with what they believe. We trust their testimony, even when we shouldn't.

My mom's skepticism about the reliability of testimony has deep roots in our culture. The seventeenth-century philosopher John Locke snarled at the idea that you can know something *merely* because someone else tells you: "I hope it will not be thought arrogance to say," he begins, "that perhaps we should make greater progress in the discovery of rational and contemplative knowledge if we sought it in the fountain . . . of things in themselves, and made use rather of our own thoughts than other men's to find it." He goes on, "The floating of other men's opinions in our brains makes us not one jot the more knowing, though they happen to be true. What in them was science is in us but opiniatrety."[8] Locke's point seems to be that real knowledge only comes from *your own personal observations, or use of your memory, logical reasoning, or so on.* Real knowers, he seems to say, are self-reliant: they drink from the fountain of "things in themselves." That is, they believe only when they have, or least can easily obtain, reasons—reasons based on personal observations and critical thinking—for one side of the given issue or another.

An emphasis on self-reliance makes sense given that Locke was a founding figure of the Enlightenment, known for celebrating the individual's political rights and autonomy. For Locke, citizens had a natural right to their property, and the government needed to be relatively standoffish with regard to how people used that property. This trumpeting of the rights of the individual had a natural epistemic correlate, call it Locke's command: *thou shalt figure things out thyself.* The sentiment is echoed again and again throughout the period. Kant, in fact, defined enlightenment partly in terms of it: as humanity's "emergence from a self-imposed immaturity"—immaturity due to lack of courage to think for oneself as opposed to going with the flow.[9] One could not even enter the British Royal Society without passing under their motto (then and now) of *nullius in verba* ("take nobody's word for it").

These sentiments were largely a reaction to an older idea— that all knowledge requires deferring to, and trusting, authority. Education in the sixteenth century was still very much a matter of mastering certain religious and classical texts, and what you knew came from those, and only those, texts. But as it became apparent that these texts were often wrong (think of Galileo's and Copernicus' discoveries, for example) the method of consulting them for knowledge came to seem naive. Thus by 1641, we find Descartes wiping away such methods with the very first sentence of his most famous book: "Some years ago I was struck by the large numbers of falsehoods that I had accepted as true in my childhood, and by the highly doubtful nature of the whole edifice that I had subsequently based on them."[10] Descartes sat down and attempted to reconstruct what he really knew—using only materials that he could find with his own mind—and he implicitly urged his reader to do the same. In other words, don't trust someone else's say-so; question authority.

This is still good advice—to a point. That's because, really, "Locke's command" is impossible to follow strictly. We can't figure *everything* out for ourselves. So, if we interpret Locke as telling us that you only really know that which you discover on your own, then you and I don't know very much. And neither did Locke. Even in Locke's time, when someone like himself could master so much about the world (and write a book titled, non-ironically, *An Essay concerning Human Understanding*), there was still a cognitive division of labor. Expertise was acknowledged and encouraged in mathematics, navigation, engineering, farming and warfare. Education systems—universities—were already centuries old by Locke's time, and the printing press, and growing literacy, were allowing more and more people to learn from the knowledge of others.

The point is partly economic. It simply isn't efficient to try and have everyone know everything—any more than it is efficient to try and get everyone to grow all their own food. After all, how much of the information that you get from your phone could you work out personally? My experiment from last summer demonstrates that the answer is: not very much. Even if we can figure some things out offline, we still consult experts and outside resources. In all these sorts of cases, we must defer to the testimony and expertise of others.

Sure, when the zombie apocalypse comes, you want to be able to stand on your own. But this isn't the zombie apocalypse. In real life, self-reliant folks get that way because of what other people have done for them. Economically self-reliant people typically have benefited from the help of others (and from those public entities that maintain the roads, maintain armies and teach the workers how to read and write). Likewise for the mythic self-reliant believer. This is the fantasy of the rugged individual judging the truth for himself without dependence on anyone else. In the TV commercial version, he is the man in the white lab coat, smoking his pipe, squinting into a microscope. But how did he get there? By learning from others, of course—from education. And while some of the information we learn this way can, at least in principle, be verified by our own reasoning and observation, the fact is that not everything we learn is like this. All of us are finite creatures with limited life spans. We can't check out all the information we rely on in any given day, let alone over a whole lifetime.

So, while information cascades, rumors and ignorance do spread like wildfire, we aren't going to give up on Google-knowing because of that. Nor should we, any more than we should give up learning from others. The lesson I glean from Mom and Locke, therefore, isn't to be an intellectual hermit. You don't have to throw your iPhone away and stop using Twitter. *What we—both as individuals and as a society—should learn from Mom and Locke is that we must be extremely careful about allowing online information acquisition—Google-knowing—to swamp other ways of knowing.* And yet that is, increasingly, precisely what we are prone to do.

Being Reasonable: Uploading Reasons

Imagine you want to buy a good apple from folks who have their own apple-sorting device. They *claim* it is great at picking out the good apples from the bad. Does that help you decide whether to buy one of their apples? Not really—even if they were to later turn out to be right. For unless you already have reason to trust them, the mere fact they say they have a reliable apple-sorter is of little use to you. And that remains the case even if they are actually right—they really do have the good apples. Analogously, where the issue isn't sorting good apples from bad but true information from false, my merely *saying* I've got good information isn't generally enough to make you want to buy it—even if I really do have a way of telling the difference between what's true and what isn't.

This tells us something important: that if we were to define knowing as only being receptive—as accurate downloading and nothing more—then we ignore something important about the human condition.

In some cases—many cases, in fact—we trade information in situations where trust has already been established to some degree. It is a spectrum. On one end of the spectrum are cases where we have a reasonably high degree of trust in the person we are asking for information: as when you talk to your trusted family doctor about your health, your spouse about your children, your professor about the material which she is professing. Of course, no human is infallible, but you ask those you trust for information because you think they have a good probability of being right—or at least more right than yourself. Farther down the spectrum is the case where you stop a passerby to ask for directions. Here too, you already have some degree of trust, since a) you have some reason to think he is local (he is walking down the street) and b) you have no reason to think he will lie. We all know that both of these conditions can be defeated, of course (sometimes it turns out that you're asking another tourist), but nonetheless, we ask directions with the reasonable expectation of getting useful information.

Yet while many of our interactions are like this, many are

farther down the spectrum still. In some cases, we may need information but have very little to go on. That's why we seek evidence to help us assess other people's opinions. When, for example, we aren't an expert on something ourselves, we seek advice from those who say they are. But if we are wise, we also get evidence of that person's expertise: references, degrees, or word of mouth. Moreover, we look for them to explain their opinions to us in ways that make sense given what we do know. We don't trust blindly—we look for reasons.

This is another place where a philosophical thought experiment can help. The seventeenth-century philosopher Thomas Hobbes postulated that governments are rational responses to the nasty, brutish and short lives we are doomed to lead in what he called the *state of nature*—a state where everyone is against everyone and no one works together to distribute resources. His thought was that those in the state of nature would face strong pressures to form a government, allowing them to coordinate and share these resources: to stop the "war of all against all." Usually, when we think about this idea, we are thinking of shelter, food and water as resources. But it is pretty clear that justified accurate information—knowledge—is a resource as well. In order to escape the state of nature, we would need to *exchange information* in a situation where, at least to begin with, we would have fairly low levels of preexisting trust. In other words, we would face what we might call the *information coordination problem*.[11]

The information coordination problem isn't just hypothetical. All societies face it, since no society can survive without its citizens trading information with one another. But how *do* we solve it? You can't just look and see the truth in my brain. What you need is *some reflectively appreciable evidence*—you are looking for a *reason* to believe that my apple sorting is reliable, so to speak. By a "reason" here, I mean a consideration in favor of believing something. Not all reasons are good ones, of course. But when we are consciously deciding what to believe, we are engaging in our capacity for reflection, or "system 2" as Kahneman calls it. We are trying to sort the true from the false. When we do so successfully, we are knowing in a different way: we are not just being receptive. We are being reflective, *responsible* believers.

A key challenge to living in the Internet of Us is not letting our super-easy access to so much information lull us into being passive receptacles for other people's opinions. Locke and Descartes may have overemphasized the role of reason in our lives. But we can't fall into the opposite error either. Knowing now is *both* faster *and* more dependent on others than Descartes or Locke would have ever imagined. If we are not careful, that can encourage in us the thought that all knowing is downloading— that all knowing is passive. That would be a serious mistake. If we want more than to be just passive, receptive knowers, we need to struggle to be *autonomous* in our thought. To do that is to believe based on *reasons you can own*—stemming from principles you would, on reflection, endorse.[12]

But if the principles we use to evaluate one another's information are forever hidden from view, they aren't of much use. In order to solve the information coordination problem we can't just live up to our *own* standards. We need to be willing to explain ourselves to one another in terms we can *both* understand.[13] It is not enough to be receptive downloaders and reflective, responsible believers. We also need to be *reasonable*.

Reasonableness isn't a matter of being polite. It has a public point. Exchanging reasons matters because it is a useful way of laying out evidence of credibility. It is why we often demand that people give us arguments for their views, reasons that they can *upload* onto our shared public workspace. We use these reasons, for good or ill, as *trust-tags*. And the converse holds as well—if I want you to trust me, I will find it useful to give you some publicly appreciable evidence for thinking of me as credible. One way to do that is to upload a reason into our shared workspace.

Public workspaces require public rules. If we are going to live together and share resources, we need people to play by at least some of the same rules. We need them to be reasonable, ethical actors. The same holds when it comes to sharing information. If that is going to work, we need people to be reflective *and* reasonable believers—*to be willing to play the game of giving and asking for reasons by rules most of us could accept were we to stop to think about it.* Only if we can hold one another accountable for following the rules can we make sense of having a fair market of information.

But how realistic is this? Digital media gives us more means for self-expression and autonomous opinion-forming than any human has ever had. But it also allows us to find support for any view, no matter how wacky. And that raises an important question: What if our digital form of life has already exposed "reason" as a naive philosopher's fantasy? What if we no longer recognize the same rules of reason?

What if it is too late to be reasonable?

3.

Fragmented Reasons: Is the Internet Making Us Less Reasonable?

The Abstract Society

"We can conceive," wrote the philosopher Karl Popper in 1946, before television, computers or iPhones, "of a society in which men practically never meet face to face—in which all business is conducted by individuals in isolation who communicate by typed letters or telegrams, and who go about in closed motor-cars. . . . Such a fictitious society might be called a completely abstract or depersonalized society."[1]

This passage is remarkable in several ways. It is certainly prescient. The idea—if not the prose—is more like something you would have found at the time in *Amazing Stories* or the adventures of Tom Swift, rather than buried in a difficult two-volume essay on democracy, fascism and knowledge. It is also very honest. Popper was, in effect, offering a warning about his own ideas concerning open societies—namely, that such societies could easily become "abstract" or "depersonalized." Popper was a fierce advocate of openness, and he saw open societies as being marked by their values: they are committed to freedom of speech and thought, equality, reasonableness and an attitude of progressive criticism. Today we might say that a society is open to the degree that it protects freedom of communication and information, imposes little government censorship and has a diverse (and diversely owned) media.[2] By this standard, a country like the United States is reasonably open (if not as open as many of us would like it to be, and becoming less so). And the Internet is a large part of this. The Web flourishes in part because it allows us unprecedented control over the sources and types of information we receive, to dip into the flow of information where and how we wish, and to extract and isolate what interests us more quickly, all in the comfort of our pajamas. It allows us to get what we want—or what we think we want—faster. And it allows us to do so without leaving our protective bubble, without sullying ourselves with the messy and inconvenient physical lives of others; it offers anonymity and friends we've never met. And that, one might think, is precisely what Popper was warning about: that with increased freedom of expression and consumption comes the risk of increased individual isolation.

Research over the last couple of decades suggests that Popper was right to be concerned.[3] But it is not clear that the only, or even the most fundamental, problem is that we are more isolated *individuals*. Really, communication is communication—even if some methods might be better for certain purposes than others. And the information technology coursing through our society's veins has given us more ways of communicating.[4] Indeed, we can hardly get away from one another: we email, we text, we tweet and soon, maybe, we'll just think to one another. But to whom do we talk, and to whom do we listen? That's the question, and the evidence suggests that we listen and talk to those in our circle, our party, our fellow travelers. We read the blogs of those we agree with, watch the cable news network that reports on the world in the way that we see it, and post and share jokes made at the expense of the "other side."[5] The real worry is not, as Popper feared, that an open digital society makes us into independent individuals living Robinson Crusoe–like on smartphone islands; the real worry is that the Internet is increasing "group polarization"—that we are becoming increasingly isolated *tribes*.

As one of the most influential thinkers about digital culture, Cass Sunstein, has noted, one reason the Internet contributes to polarization is that "repeated exposure to an extreme position, with the suggestion that many people hold that position, will predictably move those exposed, and likely predisposed, to believe in it."[6] So, with a steady diet of Fox News, conservatives will become more conservative. Liberals who only read the *Huffington Post* or the *Daily Kos* will become more liberal. And that means, Sunstein argues, that true fragmentation of the society results, "as diverse people, not originally fixed in their views and perhaps not so far apart, end up in extremely different places, simply because of what they are reading and viewing."[7]

We are getting more and more used to fragmentation now. It is reflected in our social media. Liberals tend to be friends on Facebook with other liberals, and Twitter feeds are clogged with the tweets of daily outrage: the latest news that is sure to piss your friends off as much as it did you.[8] Yet most discussions of polarization talk about the fragmentation of our *moral and political* values. That makes sense: we live in a world of Christians and Muslims, atheists and theists, Republicans and Democrats, free-marketers and socialists, etc. These differences in religious, moral and political values are how we identify one another as members of the same tribe; and they affect our behavior in all sorts of ways, from determining who gets invited to the dinner party to which candidate we'll vote for in the election.

But could it be that the Internet is helping to fragment not only our moral and religious values, but our very standards of reason? Could it be making us less reasonable?

When Fights Break Out in the Library

Let's go back to the idea of the Borgesian library discussed in chapter 1. It encompasses the world. It contains books on every subject, from politics to physics to pencil-making. But not all the books agree. And we cannot leave the library to find out which is right and which is wrong. It is all there is.

Were we to live in such a Borgesian library world—as we do, in an obvious sense—we would be in a state of *information glut.* Information on any topic we can imagine—and much that any particular individual can't—is contained within the infinite walls of the library. Some of it will be accurate and some partly so, and some complete gibberish. The question is how to tell which is which. Theories will be propagated, and certain reference books will be seen as keys to unlock the secrets of the other books or as useful maps to the truths and falsehoods. But people will disagree over which books are the best references, over the very standards for sorting the good books from the bad.

How would people react in such a situation? A natural reaction would be an increasing tribalization of the sort we saw Sunstein remarking on above. Just as in a room of shouting people you start by focusing on the voices you recognize, so the library-dwellers would be prone to read some books over others, and to discount not only their rivals' books but their reference books—the very standards they use to sort good books from bad. As such, rational discussion about whose books are best, and how to sort new books that come in, will become increasingly difficult. Tribes within the library will evaluate one another's reasons by completely different standards.

The intellectuals of the world will nod their heads sagely. It is inevitable, they will say. There is no way to know which books contain the objective truth, some will announce. "There is no objective truth!" others will assert; all books are relative to other books. Still others will declare that only faith in the one true book can solve the problem—appeals to references, citation records, card catalogues and rational standards generally is all for naught. Such reactions will only increase tribalization, and the more practical-minded of the inhabitants may begin to listen to those who point out that the only real way to settle the issue is to burn the other tribes' books.

There are reasons to think we are living in this sort of environment now. By giving us more information, the Internet not only gives us more things to disagree about, it allows us to more easily select and choose those sources that validate our existing opinions. And that, in turn, can cause our disagreements to spiral ever deeper.

Consider for example, so-called "fact-checking" sites like Politifact, which was started by the old media outlet the *St. Petersburg Times* to help cut through all the tribe-talk and verify different claims to truth made in the cultural and political debates that fill the news. And by and large, they've had a healthy impact. But they've also come under increasing assault themselves. In his essay "Lies, Damned Lies, and 'Fact-Checking': The Liberal Media's Latest Attempt to Control the Discourse," Mark Hemingway charged that fact-checkers are themselves biased—toward the left. His evidence: several examples where fact-checkers seem to get things wrong, in a politically biased way. According to Hemingway, "What's going on here should be obvious enough. With the rise of cable news and the Internet, traditional media institutions are increasingly unable to control what political rhetoric and which narratives catch fire with the public. Media fact-checking operations aren't about checking facts so much as they are about a rearguard action to keep inconvenient truths out of the conversation."[9]

Notice how Hemingway frames his fact-checking of the fact-checkers. He takes himself to be exposing a hidden truth: the truth that some folks (the fact-checkers) are keen to keep inconvenient truths out of the conversation. Whether or not Hemingway is right about his claim, the point here is that the truth wars in this country have grown to such proportions that the very idea of "fact-checking" is seen as suspect.

Once debates reach this point they are very difficult to resolve. It has become a matter of principle. Not moral principles but "epistemic" principles—"epistemic" because they are about what is rational to believe and the best sources of evidence and knowledge. Disagreements over principles such as these illustrate a very old philosophical worry: namely, that all reasons end up grounding out on something arbitrary.

For example, suppose I challenge your epistemic principle P which says that such-and-such a method is a reliable means to the truth. You defend it by appeal to some other principle, Q. If I persist in my skepticism and question Q, your options seem to dwindle. Pretty soon, being a finite creature with a finite mind, you are going to run out of principles. It seems that you must either end up defending your Q with P (whose truth is still not established) or simply dig in your heels and tell me to take off. Either way, you haven't answered my challenge, and your faith in your principles—and therefore the very methods you use to reach the truth about matters both mundane and sublime—seems blind.

This paradox goes back at least as far as the pre-Socratic philosophers of ancient Greece. Yet it reappears in cultural debates like clockwork. Today, it is heard in the claims made by evangelicals to the effect that science is really just another religion: "Everyone, scientist or not, must start their quests for knowledge with some unprovable axiom—some *a priori* belief on which they sort through experience and deduce other truths. This starting point, whatever it is, can only be accepted by faith. . . ."[10] This is a powerful idea—in part because it contains more than a grain of truth, and in part because it simply feels liberating. It levels the playing field, intellectually speaking. After all, if all reasons are grounded on something arbitrary, then why assume science is on any firmer foundation than anything else? You might as well just go with what you already accept on faith.

If we were to concentrate just on the receptivity model of knowledge that we saw in chapter 1, then such debates wouldn't threaten knowledge at all. But that misses the point. Because the problems they cause are not for receptivity but for reasonableness. What they threaten is our ability to articulate and defend our views. The problem that skepticism about reason raises is not whether I have good evidence by my principles for my principles. Presumably I do.[11] The problem is whether I can give a more objective defense of them. That is, whether I can give reasons for them that can be appreciated from what the eighteenth-century philosopher David Hume called a "common point of view"—reasons that can "move some universal principle of the human frame, and touch a string, to which all mankind have an accord and symphony."[12]

Those who wax skeptical about the use of scientific methods to resolve debates such as the origins of life on earth, or the beginning of the universe, for example, are rarely if ever skeptical about science across the board. Their quarrel is with its use in certain domains. The folks at AnswersinGenesis.org aren't going to say that we should never use observation, logic and experiment to figure things out. What they will argue is that these methods have a lower priority in some subject matters than others, where other methods trump them. People who think that the Torah or the Bible or the Koran is a better—not the only—means to the truth about the origin of our planet, for example, see the matter in that way.

Imagine a dispute between a scientist and a creationist over these two principles:

(A) Inference to the best explanation on the basis of the fossil and physical record is the *only* method for knowing about the distant past.

(B) Consultation of the Holy Book is the *best* method for knowing about the distant past.

The friends of (B) aren't rejecting outright the strategy of consulting the fossil and historical record. So we can't just call them out for using it sometimes and not others. And obviously, we can't travel back in time and use observation (another commonly shared method) to settle who is right and who isn't about the distant past. What this shows is that debates over even very specific principles like these can end up grounding out—either the participants will end up defending their favored principles by appealing to those very principles (e.g., citing the Book to defend the Book) or appealing to specific principles that the other side shares but assigns a lower priority. Neither side will be able to offer reasons that the other will recognize for his or her point of view.

As I've already noted, the Internet didn't create this problem, but it is exaggerating it. Yet you might think that this isn't so bad. As philosopher Allan Hazlett has pointed out, if everyone agrees in a democracy, something's gone wrong.[13] Democracies should be, in John Rawls' words, places where there are "a plurality of reasonable yet incompatible comprehensive doctrines."[14]

But that's just the point. How do we figure out, as a society, whose views are "reasonable" and whose are not, if our standards for what counts as reasonable don't overlap? And how do we engage in *dialogue* with people with worldviews that are different from our own (as opposed to oppressing them, or manipulating them, or simply shouting at them) without an exchange of reasons? The answer is: we don't. And that tells us something: civil societies not only need a common currency to exchange money; they also need a common currency to exchange reasons.

So, the point is not that we should all agree. We all have different experience bases, after all, and that means we can use different evidence even if we agree on what counts as evidence. But if we don't agree on what counts as evidence, on our epistemic principles, then we aren't playing by the same rules anymore. And once that happens, game over.

The Rationalist's Delusion

Perhaps we shouldn't be surprised that we have a hard time defending our "first principles" with reasons. It might be that the fragmentation of reason, while exaggerated by our digital culture, is actually the result of human psychology. If so, then perhaps being reasonable, defined as the willingness to give and ask for reasons that others can appreciate, is an untenable ideal. After all, you don't have to be Karl Rove to suspect that the evidence often fails to persuade and that what really changes opinion is good advertising, emotional associations and having the bigger stick (or super PAC).

Recently, some social scientists, most notably the psychologist Jonathan Haidt, have suggested that this is not far from the truth.[15] Haidt has done remarkable work exposing some of the psychological causes of our divisions in values. But he thinks this work shows that the philosopher's dream of reason isn't just naive, it is radically unfounded, the product of what he calls "the rationalist delusion." As he puts it, "Anyone who values truth should stop worshipping reason. We all need to take a cold hard look at the evidence and see reasoning for what it is."[16] Haidt sees two points about reasoning to be particularly important: the first concerns the relative efficacy (or lack thereof) of reasoning; the second concerns the point of doing so publicly: of exchanging reasons. According to Haidt, value judgments are less a product of rational deliberation than they are a result of intuition and emotion. In neuroscientist Drew Westen's words, the political brain is the emotional brain.

If this is right, then we not only have something of an explanation for why knowledge fragmentation continues to persist (people just won't listen to one another's reasons) but also a lesson for what to do about it. Or at least what not to do: trying to come up with reasons to convince our cultural opponents isn't going to work. If peace is in the offing, it is going to have to come about some other way.

There is, without question, a lot of sense to the idea that reasons often—perhaps mostly—fail to persuade. As we've already seen from Kahneman's work, our reflective self, whose job it is to monitor our fast judgment-making processes, is often not on the ball. And even when it is, "reasoning" often seems to be post-hoc rationalization: we tend to accept that which confirms what we already believe (psychologists call this confirmation bias). And the tendency goes beyond just politics. When people are told that they scored low on an IQ test, for example, they are more likely to read scientific articles criticizing such tests; when they score high, they are more likely to read articles that support the tests. They are more likely to favor the "evidence," in other words, that make them feel good. This is what Haidt calls the "wag the dog" illusion: reason, he says, is the tail that we mistakenly believe wags the dog of value judgment.[17]

Much of this has to do with our brain's ability to trump reason with emotion. Consider some of Haidt's own well-known research on "moral dumbfounding." Presented with a story about consensual, protected sex between an adult brother and sister—sex which is never repeated, and which is protected by birth control—most people reacted with feelings of disgust, judging that it is wrong. Yet they struggled to defend such feelings with arguments when questioned by researchers.[18] Even so, they stuck to their guns. Haidt suggests that this means that whatever reasons they could come up with seem to be just along for the ride: their feelings were doing the work of judgment.

Data like this should give us pause, but we need to be careful not to exaggerate the lessons it has to teach us. The inability of people—in particular young college students like those in Haidt's study—to be immediately articulate about why they've made an intuitive judgment doesn't necessarily show that their judgment is the outcome of a nonrational process, or even that they lack reasons for their view. Intuitions, moral or otherwise, can be the result of sources that can be rationally evaluated and calibrated.[19] Moreover, rational deliberation is not a switch to be thrown on or off. It is a process, and therefore many of its effects have to be measured over time. Tellingly, the participants in Haidt's original harmless-taboo studies had little time to deliberate. But as other studies have suggested, when people are given more time to reflect, they *can* change their beliefs to fit the evidence, even when those beliefs might initially be emotionally uncomfortable.

Haidt has been careful to say that reasons do play some role in moral and political judgments. His point is that reasons are far less influential than intuition and emotion. The latter factors trump reasons: "reasons matter (except when intuitions object)."[20]

As I've said, it is hard to argue with this—even just based on the anecdotal evidence that daily life provides. But it doesn't show that reason doesn't have a role to play. Consider the changing attitudes toward homosexuality and same-sex marriage in the United States. What caused this change? Part of the story is simply that younger people, in general, are increasingly tolerant of same-sex marriage. Another part is increased contact and exposure to gays and lesbians through the media. But the battle over same-sex marriage has also been partly a legal battle, where the issue has concerned not just the definition of marriage but the alleged harm same-sex unions cause to others not in those unions. Tellingly, the evidence—reason—has shown those claims to be unjustified.[21] And that fact seems to have had an impact on judicial proceedings about the matter—most famously in the 2009 Proposition 8 legal case, when the attorney arguing the case against same-sex marriage was reported to have conceded in court that he did not know what harm would result from letting same-sex partners marry.[22] So perhaps we can explain massive moral and political change of this sort without having to invoke the causal influence of reasons, but it seems just as likely that appeals to evidence—evidence, in fact, often uncovered by social scientists—have had an impact on how people view same-sex (or interracial) marriage via affecting institutions such as the law.

Moreover, as the psychologist Paul Bloom has pointed out, it seems likely that rational deliberation is also going to be involved in the creation of *new* moral concepts—such as human rights, or the idea that all people should be treated equally under the law.[23] Changes in moral concepts are often changes that occur despite the resistance of the "intuitive" or "emotional" judgments people inherit from the culture around them. But such changes take time. So, to show that reasons cannot trump intuition in value judgments, we would need to show that they don't change our moral judgments *over time*.

This brings us around to Haidt's second main point about reasoning, mentioned above. He endorses what he calls a Glauconian view of reasoning about value. The reference here is to an old saw from Plato: What would you do with a ring of invisibility? Fight for truth, justice and the American way—or spy and steal? In Plato's *Republic,* the character Glaucon asks this question to illustrate the idea that it is merely the fear of being caught that makes us behave, not a desire for justice. Haidt takes from this a general lesson about the value of defending our views with reasons. Just as those who do the "right" thing are not really motivated by a desire for justice, those who defend their views with reasons are not "really" after the truth. As the cognitive scientists Hugo Mercier and Dan Sperber put it, the function of both reasoning and the exchange of reasons is persuasion and persuasion alone. If so, then what people are really after when looking for reasons—whether they acknowledge it or not—are arguments supporting their already entrenched views, and/or a way to push people into agreeing with them.[24] So even if appeals to evidence are sometimes effective in changing our values over time, that's because reasons themselves are aimed at manipulating others into agreeing with us, not because they might have also uncovered the facts. On this view, to think otherwise is to once again fall into the rationalist delusion.

Anyone who has spent time on the Internet will probably feel the pull of the Glauconian view of human rationality. Social media and the blogosphere are filled with "reasoning"—but much of it seems to be either blatant marketing or aimed only at supporting what people already believe. Maybe that is what the Internet is teaching us. Maybe we are all Glauconians, and always have been.

Democracy as a Space of Reasons

I began this chapter by asking whether the Internet is making us less reasonable. Being reasonable, I've said, amounts to defending your views with reasons that are in line with shared epistemic principles or standards. We've canvassed two deep challenges to reasonableness so defined. The first stems from an ancient philosophical paradox. It points out that when disagreements go all the way down to epistemic principles, reasonableness goes by the board. The second challenge comes from results in social psychology. It forces us to wonder whether reasons are really effective tools for persuasion at all.

Neither of these challenges is new. And so it would be wrong to say that the Internet itself is making us less reasonable. It would be more accurate to say that we are making ourselves less reasonable with the help of the Internet. Or more precisely still, that the Internet is exaggerating these challenges, making them even more pressing.

In both cases, it is the very availability of so much information—our life in the library—that is part of the problem. That's a point Haidt emphasizes: "Whatever you want to believe about the causes of global warming or whether a fetus can feel pain, just Google your belief."[25] Our ability to access so much information just makes it easier than ever to follow our hardwired tendencies to make the facts fit what we already think.

There are reasons for hope, of course. We actually do use the Internet to hold one another to account—to solve the information coordination problem I talked about in the last chapter. Think of the ubiquitous smartphone check. How often have you been at a party, or in a bar, or in a lecture, and someone makes some point of fact; out come the phones and a race is on to see who can verify (or falsify) it first. We are holding one another accountable when we do this (and sometimes also being annoying). Wikipedia has become one of our most widely shared public standards of evidence. And that is often a very good thing—it cuts down on irresponsible assertions (even if it also cuts down on spontaneity). Moreover, as we'll see later in the book, there are obvious ways in which the Internet can be a force for social cohesion and democratic discussion.

We also shouldn't be too willing to accept the Glauconian view of the function of human rationality. That's partly because human rationality is too complex to have a single kind of function. In giving reasons, we certainly aim to get others to agree with us (I'm doing that now, after all). And aiming at agreement is a good thing, as is searching out effective means of reaching it (indeed, this is one of the noble ideals of Haidt's book). But it is less clear that we can coherently represent ourselves as *only* aiming to get others to agree with us in judgment.

To see this, think about the difficulty in being skeptical about the role of rationality in our lives today. The judgment that reasons play a weak role in judgment is itself a judgment. And the Glauconian skeptic has defended it with reasons. So, if those reasons persuade me of his theory despite my intuitive feelings to the contrary, then reasons *can* play a trumping role in judgment—contra the theory.

Of course, one might reasonably say that the reasons to accept views like Haidt's are not value judgments. They are scientific claims. But even the most "scientific" of claims is *informed* by value judgments. Science itself presupposes certain values: truth, objectivity and what I call epistemic principles—principles that give us our standards of rationality. Moreover, outside of mathematics it is rare that the data is so conclusive that there is just one conclusion we can draw.[26] Usually the data admits of more than one interpretation, more than one explanation. And that means that we must infer, or judge, what we think is the case. And where there is judgment, there are values in the background. Hence the point: arguing (with reasons) that reason never plays a role in value judgments is apt to be self-defeating.

There is a larger point here. Even if we *could* start seeing ourselves as only giving reasons to manipulate, it is unclear that we *should*. Suppose I offered you a drug that, once dropped in the water supply, would make most folks agree with your political views. It would be tempting, wouldn't it? But it would also be wrong. And it is wrong in the very way that we think the sleaziest political ads are wrong. To engage in democratic politics means seeing your fellow citizens as equal autonomous agents capable of making up their own minds. That means that in a functioning democracy, we owe each other reasons for our political actions. And obviously these reasons can't be "reasons" of force and manipulation, for to impose a view on someone is to fail to treat her as an autonomous equal. That is the problem with coming to see ourselves as more like Glauconian rhetoricians than reasoners. Glauconians are marketers; persuasion is the game and truth is beside the point. But once we begin to see ourselves—and everyone else—in this way, we cease to see one another as equal participants in the democratic enterprise. We are only pieces to be manipulated on the board.

This ethical or political point doesn't, of course, settle the issue of whether our psychologies are hardwired in such a way that reasons have very little influence over us. But it does illustrate what's at stake, and cautions against drawing conclusions too quickly.

A similar point can be made about the first, and older challenge to reasonableness stemming from the ancient skeptics, as another story from the history of philosophy suggests.

Johann Friedrich Zöllner was an eighteenth-century clergyman and political essayist now remembered for being the guy who inspired Kant to define the Enlightenment. The context will sound weirdly familiar. At the beginning of the French Revolution, Zöllner published an article in the *Berlin Monthly* opposing the progressives of his day, who were arguing that marriage should be treated as a civil, not a religious institution. Zöllner said that only religion could provide a proper basis for marriage and that religious authorities should be given more weight in civil matters. Enlightenment values, as the progressives called them, weren't up to snuff. And besides, he sneered in a footnote, no one could ever explain what "enlightenment" meant anyway.

A few months later, Kant published a direct answer to Zöllner's challenge. We encountered it in the first chapter: enlightenment, Kant said, means having the courage to think for yourself. Thus the Kantian bumper sticker: *Sapere aude*; dare to know.

Kant's concern was partly with intellectual autonomy. But he also points out that we are beings that *can* think for ourselves, and so in our role as citizens we owe it to one another to explain ourselves in ways that respect that fact. That's because, Kant says, when I give you reasons I treat you as someone who is free to make up your own mind. I treat you with dignity. I treat you as a grownup. So, even if you really do *know* the truth—even if you are an oracle with all-knowing powers, or Plato's philosopher king—you shouldn't just appeal to that fact in public debate. We owe one another reasons that appeal to our shared humanity— that others have the potential to recognize *as* reasons just because they are human.

Kant's point helps to mitigate the force of the ancient skeptical argument even if it doesn't answer it directly. The skeptical argument says, in effect, that we can't defend fundamental scientific methods as any more rational than other methods. What Kant points out, however, is that we *can* show that they are more democratic, more respectful of basic human autonomy. Why? Because scientific methods use human cognitive capacities such as observation and inference. That doesn't mean these capacities are always reliable, or even that we are very skilled at using them (news flash: we aren't). But human capacities like these—capacities that are at the very basis of science—do have an obvious virtue for a digitalized society: they aren't secret or the province of a few. Observation and logic are strategies that everyone can, at least to some extent, use themselves and employ in their social networks, and that can be made at least a little more effective with training. It is no coincidence that Locke and other champions of science were also champions of what we now call human rights. Prioritizing scientific methods is liberating precisely because it frees one from appeals to authority, from the thought that something is true because someone in power says so.

The Internet has created an explosion of what I called in the last chapter receptive knowledge. We saw there that while this is wonderful in many respects, it isn't enough; we need to exchange reasons and play by shared epistemic rules if we are going to solve the information coordination problem that faces all societies. But Kant reminds us that reasonableness defined in this way also has serious political and democratic value. That's why it is so crucial that we pay attention to how we are encouraging people to know about the world, and in particular the sorts of institutions that help them do that. As Haidt has remarked, we aren't going to get people to be more reasonable with one another by having them sign "civility pledges."[27] We need to promote institutions that encourage cooperation, and even face-to-face contact with people who have very different views. And more than that, we need to promote institutions that encourage us to engage our capacities for receptive thought. Institutional structures can

60

help us overcome our private limitations—our biases, implicit or otherwise. That is what institutions are for. And that's why, even if in private life you think of the Bible, or the Koran or *Dianetics*, as the ultimate authority on the universe, in *public* life you should support institutions like the National Science Foundation or the National Endowment for the Humanities or, frankly, your local university. Institutions that encourage the use of critical thinking and the civil exchange of reasons are *doing the work of democracy*. In part, that is because broadly scientific principles of reasonableness privilege principles that everyone appeals to most of the time—just because we are built that way.

Indeed, that's the very reason some people don't like the idea that we should privilege these sorts of principles in public discourse. Consider this little item from the "you can't make this up" department. In 2012, the Republican Party in Texas included the following in their platform:

Knowledge-Based Education – We oppose the teaching of Higher Order Thinking Skills (HOTS) (values clarification), critical thinking skills and similar programs that are simply a relabeling of Outcome-Based Education (OBE) (mastery learning) which focus on behavior modification and have the purpose of challenging the student's fixed beliefs and undermining parental authority.

Part of this is inside-baseball education-speak: the real target is "Outcome Based Education," not critical thinking per se. The interesting point here is why they are opposed: because it challenges the student's fixed beliefs and undermines authority.

I have no opinions on "Outcome-Based Education," nor do I think that all Republicans are against critical thinking. My point is that this particular critique is profoundly, utterly undemocratic. It also illustrates Kant's point. We need to privilege "scientific" epistemic principles and methods of thinking in public discourse precisely because such principles allow us to *evaluate* authority. What makes scientific methods of rationality important is that without them you can't hope to have anything like an open society. Critical thinking—the teaching of it, and the use of it in political argument on the Web and in the media at large—matters because without it, we fragment.

The philosopher Richard Rorty famously declared that, "if you take care of freedom, truth will take of itself." His idea, which he found in the educator and philosopher John Dewey, was that we can't hope to ground our political principles on our scientific or epistemic principles. We can't hope for a "foundationalist" view that places science on the bottom, holding up democracy. That's because sometimes it goes the other way around: we have to ground our fundamental epistemic principles on our democratic values. But that doesn't mean we should put politics first, science and epistemology second. Foundationalism turned on its head is still foundationalism. The right lesson to draw is one Kant would have thought obvious: our political and intellectual values are intertwined. The hard part isn't seeing this fact; it is in trying to make sense of how we should improve our values—epistemic, intellectual and political—making sure that truth and freedom take care of each other.

However we ultimately solve this problem, accelerating fragmentation is not to be taken lightly. Civil society requires that we treat one another with respect. We need to view one another (at least some of the time) as autonomous thinkers—as persons who can make up their own mind and have the right to do so. Give up on that and you give up on a central element of what Dewey would have called the public life: a common currency of principles and reasons that we can use to sort information and disputes over that information. The worry we've canvassed in this chapter is that the infosphere is making a true public life harder to achieve. We live in a Library of Babel, isolated in our separate rooms, poring over information culled from sources that reinforce our prejudices and never challenge our basic assumptions. No wonder that—as the debates over evolution, or what to include in textbooks, illustrate—we so often fail to reach agreement over the history and physical structure of the world itself. No wonder joint action grinds to a halt. When you can't agree on your principles of evidence and rationality, you can't agree on the facts. And if you can't agree on the facts, you can hardly agree on what to do in the face of the facts, and that just increases tribalization, and so on and on in a recurring loop.

Before you know it, the library has burned to the ground.

4.

Truth, Lies and Social Media

Deleting the Truth

In July of 2012, Western news outlets reported that "truth" was deleted from the Internet in China. According to Chinese bloggers, searches for the Chinese character for "the truth" on the popular Twitter-like social media site Weibo resulted in the following message: "According to relevant laws, regulations and policies, search results for 'the truth' cannot be displayed."[1]

If the reports are to be taken seriously, Chinese censors were not content with preventing people from accessing truth; they wanted to prevent people from even discussing what it is.

Over and above its inadvertent hilarity, this reminds us that the Internet is a revolutionary tool partly because it allows people to look for the truth on their own—independently of what governments, the scientific establishment or their own mother think is true. Perhaps the most striking example of this was the

Arab Spring. As is now well documented, social media—specifically Twitter—not only allowed protesters to effectively organize, it gave them a way to let the world know about what was
happening in their countries—countries that were ruthless in
squashing regular media outlets. Since then, Twitter activism
has only increased, and protest movements around the world use
it to get their message out and to speak truth to power. That's
widely known.

It is *also* widely known that the Internet is the world's most
powerful tool for controlling and distorting the truth. We've
already talked about how the geography of the Internet encourages herd mentalities, lemming-like information cascades and
group polarization. But it is also, in the philosopher and critic
Jason Stanley's term, an excellent "vehicle" for propaganda.[2] And
not just in China, obviously. Ask Google, "What happened to the
dinosaurs?" and you may get at the top a framed answer "card,"
as I recently did, that says, "The Bible gives us a framework for
explaining dinosaurs in terms of thousands of years of history,
including the mystery of when they lived and what happened to
them. Dinosaurs are used more than almost anything else to
indoctrinate children and adults in the idea of millions of years
of earth history."[3] This alarming result shows that Google can,
and often is, gamed. A savvy organization can make smart use
of its metrics to get its result right on the cards themselves—so
it becomes the very first result, framed in a way that makes it
seem to the casual Google-knower to be a "fact." And the organization can do all that even while insisting—as all good propaganda does—that it isn't propaganda at all.

In short, the Internet is very much a bloody, messy battleground for the truth wars. As such, it can seem harder than ever to know what is true, and this in turn has caused some to think that the concepts of truth and objectivity have outworn their welcome.

The Real as Virtual

The problem of distinguishing the real from the unreal, or the true from the untrue, is hardly the result of the digital age. What's new is how the problem manifests itself.

Take a coin out of your pocket and hold it in your hand before you. Now look at the coin: what shape does it look like? If, like most people, you say "round," then I suggest you look again. Unless you are holding the coin directly in front of your face, chances are you are seeing a more elliptical shape. This is confirmed if you make a realistic drawing of the coin. A child might draw a circle, but a more skilled artist would draw the ellipse. Why? Because that's what we are perceiving. But if so, then we have a puzzle. The coin is circular. What we perceive is not circular. Therefore, what we perceive is not the coin.

This is the sort of argument that persuaded Locke to hypothesize that what we directly perceive are not the objects themselves but our perceptions or representations of them: our "ideas" of them, as he put it. Locke argued that this was the only way to explain how we sometimes get the world wrong. Optical illusions (like the shape of the coin) are one example.

Locke also used the "idea" idea to explain the fact that our perceptions are often relative. Here's another of his experiments, one which you may have done as a child. Take three bowls of water, one hot, one cold and one lukewarm. Put your right hand in the cold, your left in the hot, and then put both in the lukewarm. We know the result: the middle bowl will feel hot to the hand that was in the cold water and cold to the hand that was in the hot water. So, what is it that we perceive? Locke's answer, following Galileo, was that all substances had two types of "qualities." Their primary qualities were those aspects that were really "in" the objects, as Locke put it—those properties that they had independently of anyone perceiving them. Locke's favorite examples were size, shape and extension in space, but today we might say that mass is the prime primary quality. Secondary qualities, on the other hand, were not "in" the object. Instead, Locke said, they were really just the powers that objects had, by virtue of their microstructure, to cause in us certain perceptions or ideas. Colors, smells, tastes and feelings like warmth and cold were like this, he said. Thus to say a fire engine is red is not to say that it has some inherent redness in it: there are no "red" particles that compose it. Redness and other "secondary" qualities are defined by *reference to how we perceive them.*

Locke believed that all knowledge is mediated through our perceptions. Our perceptions are like goggles permanently strapped to our head. Sometimes, when our vision is clear, our perception represents the world as it really is. For Locke, this meant that our ideas are caused to actually resemble the objects outside our minds. But of course, our vision *isn't* always clear—which raises an obvious question, one made famous by Locke's contemporary George Berkeley. If we are always trapped within our perceptions, how can we ever tell which of those perceptions reflect things as they really are and which are the products of our own minds? No amount of careful checking and experiment will help, noted Berkeley, for according to Locke's own view, we can't step outside of our perceptions. We can't assume the view from nowhere.

Whether or not we agree with the details of Locke's philosophy today, it is clear that Berkeley's challenge is not going away. Indeed, in some ways, it seems more difficult than ever to tell the difference between what is real and what is subjective.

One reason for this is that the Internet is a *construction*. That's partly because the World Wide Web is obviously something we've made—a literal artifact. By a literal artifact, I mean something that has been intentionally brought into being by human activity directed at that very purpose. The servers, cables, and circuits that compose the physical backbone of the Internet are all literal artifacts. But so are the packets of information that compose the body of the Internet itself. Websites, user interfaces, jpeg files, movies on Netflix, cookies, are also literal artifacts. They are, in a clear sense, as real as anything is. But the way in which informational objects are real, and the manner in which they are constructed, is very different. That's because, as the philosopher Luciano Floridi has noted, informational objects are abstracted in that they are "typified." To talk about a music file is to talk about an object that isn't identical to any of its "tokens." That's because digital copies are indistinguishable; you can create many "copies" of a digital file simultaneously. Being all born at the same time, as it were, each is in one sense not a "copy" at all—each has equal claim to being the original.

Floridi takes this to mean that we have expanded our ordinary conception of what is real, from what is material (something you can kick) to include "objects and processes that are *dephysicalized* in the sense that they tend to be seen as support-independent."[4] But actually, we were familiar with dephysicalized objects prior to the digital revolution. After all, what is a piece of music itself? Beethoven's Fifth or Jay Z's "New York" are more abstract than even the digital files that encode them: destroy all the computers and someone could still hum the tune. Yet even if we haven't expanded our conception of what is real per se, it is clear that we've constructed some new kinds of real things. And given that our digital form of life is composed of these kinds of things, we might say that our digital form of life is literally constructed in a deeply different way than previous forms of human life.

One consequence of this fact is that it is harder to see what is and isn't constructed. Locke's view of the world requires a distinction between what is "out there" (primary qualities) and what is at least partially dependent on us (secondary qualities). But as Floridi has noted, the division between "onlife" and "offlife" is increasingly difficult to make out. Watches, glasses and phones are no longer mechanical things. They are gateways to the Internet. The difference between "brick and mortar" stores and online merchandising is similarly blurred. Already you can go into a "smart" dressing room that remembers who you are, and which has a touchscreen mirror. Thermostats, refrigerators, children's toys, tools and washing machines can be (and are) connected digitally to the Web, sending and receiving information, emails, locations, updates. This is the Internet of Things. As Floridi notes, "With interfaces becoming progressively less visible, the threshold between *here (analogue, carbon-based offline)* and *there (digital, silicon-based, online)* is fast becoming blurred, although this is as much to the advantage of *there* as it is to *here.*'"[5]

But the blurring of the distinction between online and offline isn't just due to things like smart watches. For the Internet is not just the Internet of *Things*. It is also composed of social artifacts. And these emerging social constructs are intertwined with the literal constructs. One comes along with the other.

Here's what I mean. Not all literal artifacts are social artifacts. Humans make bullets and bombs, but they aren't an inherently social kind of thing (quite the opposite, in fact). Whether something is a gun, or a chair, depends on whether it serves a certain function, and each has been invented by humans to serve those functions (to kill something in the one case, to provide a seat in the other). But those functions aren't themselves necessarily defined in terms of social factors, institutions or the like. Social artifacts, on the other hand, are partly constituted by, and therefore defined in terms of, social practices.[6] These social practices can be regulated or unregulated. Unregulated: being cool. Being cool is something that is generally cool to be. But what makes something cool is a matter of how people are perceived against a social matrix—that is, against a whole host of expectations and assumptions about how people "should" act or dress. Coolness is constructed, and constructed by social expectations. And yet being cool (or not) is a difference that can make a difference—to people's happiness in social groups. The same is the case, obviously, with more regulated social roles. What constitutes my being a husband is defined in terms of social factors: I'm a husband because I meet certain legally defined expectations and institutions. You can't define "husband" and "wife" without referring to these legal conditions. And yet again, whether one gets to be a husband or not—whether you can marry someone of the same sex, for example—is something that matters to us, that seems as real and as important as anything else. The same could be said for economies, markets, governments, money, professions, religions, laws. Such things not only would have not come about without social practices, they are constituted by social practices; you can't define them without referring to various structured ways humans have of doing things. And in each case, they matter, and we accordingly treat them as part of our reality just as much as we do the concrete parts of the world.

Social constructs can change. When they do, the changes can take us by surprise or go against our preconceptions. That's because our concepts of socially constructed artifacts, like any concept, can become embedded—or perceived as being indispensable for explaining reality. The concepts of race and gender were traditionally embedded in this way, and for many people still are. Thus, when we change how we think about them, we upset expectations and prejudices.

In the digital world, literal artifacts and social artifacts are being created in a feedback loop. Life in the infosphere is both changing and being formed by certain social constructs, and these social constructs are themselves the result of life in the infosphere. Some of these changes are matters of degree (the expansion of "friend" to include "Facebook friend"). Other changes are more radical and are generally the result of changes in expectations and assumptions—changes that are themselves brought on by changes in technology. An obvious example is our concept of property. What is it to "own" music (or writing) now, or anything that is put on the Web? Should anything on the Web be open to sharing without compensation to its original creator? (One might think this question itself is outmoded—based, after all, on older assumptions and expectations.) The concept of property is a social construction par excellence, but it is one that is very much in transition. Another example—and one that we will talk about in the next chapter—is the nature of privacy. Entering the brave new world of the Internet of Us, we are quickly becoming used to having less and less privacy, and that is changing how we understand privacy itself.

Our digital form of life is changing even our identities and how we shape them. That's relevant here because some aspects of our identity are clearly social artifacts. Our identities in the psychological sense involve a number of factors, including, as the philosopher Owen Flanagan puts it, an "integrated system of past and present identifications, desires, commitments, aspirations, beliefs, dispositions, temperamental traits, roles, acts, and actional patterns."[7] Together, these aspects form what we mean when we talk about our "self."

Typically, we think of the factors that make up the self as vulnerable to influences from the "outside"—including what other people say and do in your company. Who you are is a product of who others are, and vice versa. But how we think about ourselves, how we self-represent, matters to who we are. How you define yourself at a certain time—kind, clever, embarrassed, etc.—can come into conflict with other self-representations, and that conflict can initiate change in your overall self-conception. What's more, your self-representation can change how you react in future situations, which can itself loop back to further self-representations, and so on. It is these facts that have suggested to philosophers such as Flanagan and Daniel Dennett that the self is not just a construction, but a narrative construction. I am the product of the story I and others tell about myself, whether I know that or not.

If that is so, then we are stories that are increasingly constructed online in social networks. For an increasingly large number of people, particularly people born after the mid-nineties, who and what you are is partly defined by your online activities. In the early days of Web 2.0, this fact was less appreciated: people would post pictures of themselves doing things (drinking, partying) that would later cause them embarrassment or the loss of a job opportunity. But people are more aware of this now: a college student can pay to have his online identity "scrubbed" so as to appear more respectable—more like he might aspire to be in his own self-representations. We are conscious of the stories we tell each other and ourselves.

Online identity creation is interesting in its own right. But it is also a particularly useful example of how our digital form of life is constructed. And that fact might in turn seem to be the final nail in the coffin of objectivity. If our digital form of life obscures the very difference between what is primary and secondary, between what is made and what is found, *even in the case of ourselves*, then what is the point of talking about objectivity and truth? If the real is virtual, then how important can truth be?

Interlude: To SIM or Not to SIM

Suppose that in the future you can choose between living as a SIM and continuing your current life. Once you make the choice, the company helpfully makes you forget that you ever made a choice at all. Scenarios like this are the fodder of science fiction (Philip K. Dick alone is responsible for several books on this theme), but they've also been used by philosophers.[8] We can use them here to investigate how much we still value the idea of truth.

Imagine again that in the future, computers are able to run programs that create entire SIM worlds, indistinguishable "from the inside" from real ones. Suppose that these super-engineers travel back in time and offer you three choices (maybe they are *Matrix* fans and so offer you different color pills). They warn you that once you make the choice, you can't go back: it is a permanent long-term deal.

Choice 1 is to continue with your life just as it is now. Your friends are friendly and your lovers love you (or not, as the case may be). Choice 2 is to live the exact same life you are living now but as a SIM. How they do this is up to them—perhaps they "transplant" your brain patterns into a SIM, or perhaps they keep your body alive and just make your actual brain experience a SIM-life. Either way, they'll fix it so you don't know you are living in a SIM world, but you will be. Choice 3 is just like choice 2, with one very important exception: here some of your friends and lovers really despise you. *But you will never discover that fact*, nor will you remember ever having that information; their deceit will be perfect. From the inside, all three lives will be indistinguishable; where the first causes you joy, the others do as well; where the first causes you pain, the others do as well, and so on to the grave.

Not much of a choice, really. Forced to choose, almost all of us will prefer the first life over the second two. Perhaps some may be ambivalent; they'll flip a coin. Presumably no one will *actively* prefer the third over the first, since it involves a double deception. Either way, our reaction tells us something about how deeply we dislike deception. If you are ambivalent, then deception matters less to you than it does to others. One door is as good as another. Others of us, however, will find this attitude odd, even repugnant. We don't want just *to seem* to have friends and lovers, we want actual friends and lovers, even if there were no discernible experiential difference between the one case and the other. Moreover, we want to want to be that way: we care about not being deceived. We would no more wish to be ambivalent about which of these choices to make than we would wish to willingly enter into a deception.

Our attitude toward such choices also tells us something about our attitudes toward truth. The fact that we prefer not to be deceived—even when the deception is undetectable—suggests that our preference for believing whatever is true over not doing so remains even when it would have no effect on how we experience life.[9]

But—you might be eager to ask—what if the super-scientists offer you the option of a SIM life that is all you've ever dreamed? In this SIM life, you can be whatever you want (famous athlete, successful novelist, rock star, all three, etc.). It would be like living in a video game. Call this choice 4.

How attractive choice 4 appears to people depends on how you frame it. I've found that answers vary depending on whether we are told we can just "try it" for a short time or whether it is a permanent choice.[10] Most people would be willing to try out a SIM super-life, especially if there were no negative consequences (just as most people would try certain drugs if there were no negative consequences). Some would choose it eagerly, and for longer times—especially if their "real" life is filled with pain. But most of us would still be wary if the choice was permanent. It would be a pleasurable experience, but it would be a bit like living life on an artificial high. Nothing would be earned, and the "knowledge" and "wisdom" we'd gain over our SIM life would be figments of a computer generation. That suggests that while truth is hardly our only value, we still value it overall.

But *whose* truth, exactly?

Falsehood, Fakes and the Noble Lie

When the monologist Mike Daisey got up in front of an audience at Georgetown University in 2012, he was in the midst of a media firestorm. Daisey was the author of a brilliant, funny and very biting show called *The Agony and Ecstasy of Steve Jobs*. In the show, which has no official script, Daisey—a self-confessed techno-geek of epic proportions—describes his awakening to the facts about the production of the Apple products he loves so much, in particular his iPhone. He talks about how, posing as a businessman, he was able to get into the plant in China where all such phones are made—an operation of staggeringly immense and dehumanizing size. He gives graphic details about the conditions in which the Chinese factory laborers work, building and tortuously assembling each phone, and most of its component parts, by hand.

Daisey's show was affecting. It demonstrated just how willfully ignorant most of us are about how our digital toys are made. It provided a look behind the wizard's curtain. But it also contained some falsehoods. A few months before the Georgetown speech, Daisey had given a performance of the monologue on NPR's popular *This American Life* radio broadcast. The producers asked whether the show's claims about conditions in the Chinese plants would live up to journalistic standards. Daisey said they would. But subsequent investigation by reporters turned up inaccuracies: Daisey said he'd talked to people that his translator said he had not spoken to, for example, and that he'd visited places and personally witnessed certain events that he had not. The events in question weren't large in scale—they were more on the order of small details. But the errors were enough to prompt *This American Life* to do an entire "retraction" episode.

In his subsequent Georgetown speech, Daisey admitted that he had misrepresented some facts about what he had seen, and that he had collapsed others, so as to better present them in a dramatic form. He had, in short, taken quite a bit of narrative license. He apologized for misleading Ira Glass at NPR, and for misleading the people who had listened with the expectation that the show was a piece of journalism. But he did not apologize for making the piece. Indeed, he asserted that his point in making it was precisely to expose what is indeed an extremely important truth—a moral truth that had been largely hidden from consumers. He was sacrificing certain small truths in order to expose one big truth.

The Mike Daisey story fascinates for lots of reasons. It has all the makings of a classic tragedy: a hero in pursuit of a noble truth, brought down by hubris; a touching and philosophical speech in the denouement. But it is also interesting precisely because it isn't isolated. A common theme, often voiced by the person caught faking details, is the "sacrificing small truths for one big truth" idea. And folks aren't always apologetic. The writer John D'Agata even goes so far as to suggest, in *The Lifespan of a Fact*, that the distinction between fiction and nonfiction is illusory. D'Agata, in writing what the rest of us call nonfiction articles, attempts, he says, "to reconstruct details in a way that makes them feel significant even if that significance is one that doesn't naturally occur in the event being described. . . . I am seeking truth here, but not necessarily accuracy."[11]

Now, clearly, any attempt to tell something complicated in a narrative way—the story of the Civil War, for example, as told by a documentary—will necessarily sacrifice some detail in order to illustrate the sweep of events. And in skipping some details, as any description of an event must do, it could be thought to "sacrifice" truth in a certain respect. But these facts don't bother most people. People know that details must often be left out of historical narratives. And they know that *recreations* of historical events must "fill in" where there is imperfect knowledge of the past. As a result, we adjust our expectations and, if we are wise, guard against taking the recreation as anything other than a bit of historically informed fiction.

Daisey's case, however, illustrates how quickly these expectations can shift. We typically don't treat storytellers as journalists—and this expectation didn't necessarily change just because Daisey was on NPR. It changed because of his message. His message was directed at uncovering a significant hidden truth about how iPhones were made—a truth that he thought, correctly, needed to be brought to light. That is, what his show was *about* was the fact that we were ignoring certain facts. When that is the content of your message, expectations change. We expect you to give us the right details, or to explain to us more clearly—as Daisey now does during the show—how and when details are being sacrificed for narrative drive. And this expectation rises the more we are convinced that there is something important at stake. That's why D'Agata's "I'm after truth, not accuracy" excuse rings hollow. If, as D'Agata says he is doing, I tell you that something is "significant," I shift your expectations. So you are right to feel offended if I then ignore the very expectations I've helped create. When people sacrifice small-"t" truth—what D'Agata calls accuracy—for one big Truth, their deceptions essentially involve manipulating our expectations.

The Internet makes it ridiculously easy to manipulate people's expectations. That's partly due to the relative anonymity of the Internet. But it is also because the expectation-setting context is increasingly difficult to track.

Sock puppets are a good example. A "sock puppet" is Internet-speak for a manufactured online identity used to get people to believe information of some sort. A hotel manager or restaurant owner who logs on to a consumer review site to review his own business, or hires other people to do so, is using sock puppets. Writers who review their own work under aliases on Amazon, and attack the work of others under that same alias, are doing the same. At a more sinister level, governments have been known to use sock puppets and social media to influence public opinion. Westerners typically associate such behavior with China and other oppressive regimes, and for good reason. In 2013, China was widely accused of creating dozens, perhaps hundreds, of fake Twitter accounts for propaganda purposes with regard to Chinese–Tibetan relations. But Western governments are hardly shy about using sock puppets. As the *Guardian* newspaper reported in 2011, the U.S. government has created Operation Earnest Voice, which awarded the Californian company Ntrepid millions of dollars to create sock puppets for the explicit purpose of spreading propaganda on the net in languages such as Arabic.[12]

One common form of sock puppetry is the use of a socialbot. A socialbot is not one person pretending to be another (or many), but a robot pretending to be a human. By "robot" I don't mean a walking-talking robot of the sci-fi variety; I mean an algorithm-guided bit of software that steers its false human face to like real people's sites, to make posts, and to get others to like those posts. Socialbots have been amazingly good at fooling people. They independently post and repost, tweet and retweet about current events—all using expanding databases of information gleaned from the Internet. They can respond to emails. They are often programmed to tweet in patterns that mimic awake/sleep cycles. In one famous case, a well-known Brazilian journalist—allegedly with more online influence than Oprah—was revealed to be a bot.[13] Alan Turing claimed sixty years ago that if a machine could durably fool humans into thinking it was human, then we had as much reason to think it was thinking as we have to think other human beings are thinking. By some standards, bots might seem to be passing this test.

Even if we don't think they are thinking (and I don't), the use of bots is incredibly disturbing. Part of the reason is that they are so cheap—you can buy "armies" of them for just a few hundred dollars. But they are also massive deceit machines, built for the purpose of getting people to buy things, do things, vote for certain candidates and not others. (This is partly why Twitter recently banned such bots.) Again, not all uses of bots may be harmful. But it is a mistake to write the whole technology off as simply a new and updated form of marketing or advertising. (Excusing one's manipulative behavior by saying it is "just advertising" is a bit like excusing one's infidelity as "just flirting.") In reality, these bots are more like cons. They operate on getting people to assume they are dealing with someone real who is sincere in their assertions. And they take advantage of that.

As in the Daisey case, some people use their online personas, or bots, to try and get across general political or moral viewpoints (what they consider "big truths") while perhaps sacrificing or ignoring inconvenient details. In many cases, this might be inconsequential, or simply a form of self-promotion. But there may be more at stake.

In recent years, various political organizations have used social media to great effect for propaganda. Again, the idea is often to broadcast what is perceived to be a "big truth" by misrepresenting the facts. One particular method is the use of photo-sharing. A common technique is to use an older photo but represent it as having been taken during a more recent event. For example, a widely circulated photo on the Internet—which was reproduced by various reliable news services—showed a young child jumping over rows of covered corpses. The photo was represented as having been taken in the aftermath of the Houla massacre in Syria in 2012. In fact, the photo was taken a decade earlier in Iraq. Similar uses of photos have been widely documented, and have led to several efforts to come up with verification techniques to help the public and journalists spot such abuses.[14]

In *The Republic*, Plato explicitly suggests that it would be good for citizens to believe a myth that would have the effect of making them care about their society and be content in keeping it stratified. He called this "the noble lie," and he seemed to think that it might be inevitable if the state is to survive. And of course, deception sometimes *is* justified: to protect someone, or to prevent a panic, or to minimize offense. Life is complicated and moral principles must always be applied with a sense of context. But the problem with so-called noble lies is that they are like potato chips: it is hard to stop with just one. That's because the moment a noble lie is discovered to be just that—a lie—it suddenly becomes just as "noble" to lie about whether it was really known to be a lie. Cover-ups become noble lies. Assassinations become noble lies. And soon we are sliding down the slope of the deck right into the jaws of the shark.

Objectivity and Our Constructed World

Life in the Internet of Us can make it hard to know what is true. That's partly because the digital world is a constructed world, but one constructed by a gazillion hands, all using different plans. And it is partly because it seems increasingly difficult to step outside of our constructed reality.

These facts have led some to claim that objectivity is dead. Internet theorist David Weinberger, for example, has suggested that objectivity has fallen so far "out of favor in our culture" that the Professional Journalists' Code of Ethics dropped it as an official value (in 1996, no less). Weinberger himself argues that our digital form of life undermines the importance of objectivity, in part because humans always "understand their world from a particular viewpoint." That's a problem, apparently, because objectivity rests on a metaphysical assumption: "Objectivity makes the promise to the reader that the [news] report shows the world as it is by getting rid of (or at least minimizing) the individual, subjective elements, providing, [in the philosopher Thomas Nagel's words], 'the view from nowhere.'"[15] For Weinberger, objectivity is an illusion because there is no such thing as a view from nowhere.

Maybe not. But that doesn't rule out objectivity—since being objective doesn't require a view from nowhere. Truths are objective when what makes them true isn't just up to us, when they aren't constructed. But a *person* is objective, or has an objective attitude, to the extent to which he or she is sensitive to reasons. Being sensitive to reasons involves being aware of your own limitations, being alert to the fact that some of what you believe may not be coming from reasons but from your own prejudices, your own viewpoint alone. Objectivity requires open-mindedness. It doesn't require being sensitive to reasons that (impossibly) can be assessed from no point of view. It means being sensitive to reasons that can be assessed from multiple and diverse points of view. Being objective in this sense may not always, as Locke would have acknowledged, necessarily bring us closer to the real truth of the matter, to what Kant called "things in themselves." But that is just to repeat what we already know: being objective, or sensitive to reasons, is no guarantee of certainty.

In Weinberger's view, objectivity "arose as a public value largely as a way of addressing a limitation of paper as a medium for knowledge."[16] The need to be objective, he argues, stems from the fact that paper—he means the printed word, roughly—is a static medium; it forces you to "include everything that the reader needs to understand a topic." In his view, the Internet has replaced the value of objectivity with transparency—in two ways. First, the Internet makes it easy to look up a writer's viewpoint, because you can most likely find a host of information about that writer. And second, it makes a writer's sources more open: a hyperlink can take you right to them, allowing you to check them out for yourself.

I agree that transparency in both of these senses is a value. And our digital form of life has, indeed, increased transparency in some ways—but not in all. It has increased transparency for those who already desire and value it. But as the use of sock puppets and bots demonstrates, the ability of the Internet to allow deceptive communication leads in precisely the opposite direction. Moreover, we value being able to check on sources and background precisely because we value objective reasons—reasons that are not reasons only for an individual but are valid for diverse viewpoints. We want to know whether the source of our information is biased because we want to sort out that bias from those facts that we can appreciate independently of bias. Transparency is not a replacement for valuing objectivity; it is valuable because we value objectivity.

Just as we can't let the maze of our digital life convince us to give up on objectivity and reasons, it shouldn't lead us to think that all truth is constructed, that truth is whatever passes for truth in our community, online or off. What passes for truth in a community can be shaped all too easily. That's what the noble lie is all about. What passes for truth is vulnerable to the manipulations of power. So, if truth is only what passes for truth, then those who disagree with the consensus are—by definition—not capable of speaking the truth. It's no surprise, then, that the idea that truth is constructed by opinion has been the favorite of the powerful. In the immortal words of a senior Bush advisor, "We are an empire now, and when we act, we create our own reality."[17]

That, fundamentally, is why we should hold onto the idea that at least *some* truth is not constructed by us—even if the digital world in which we live is. Give up that thought, and we undermine our ability to engage in social criticism: to think beyond the consensus, to see what is really there.

//Part II.

How We
Know Now

5.

Who Wants to Know: Privacy and Autonomy

Life in the Panopticon

In 1890, Samuel Warren and Louis Brandeis published an article in the *Harvard Law Review* arguing for what they dubbed "the right to privacy." It made a splash, and is now one of the most widely cited legal articles in U.S. history. What is less known is what precipitated the article. The Kodak camera had just been invented, and it (and cameras like it) was being used to photograph celebrities in unflattering situations. Because of this newfangled invention, Warren and Brandeis worried that technology—and our unfettered use of it—was negatively affecting the individual's right to control access to private information. Technology seemed to be outstripping our sense of how to use it ethically.

They had no idea.

In the first part of this book, we've seen how some ancient philosophical challenges have become new again. We've grappled with whether "reasonableness" is reasonable and whether truth is a fantasy. But these old problems are only half the story. To really appreciate how we can know more but understand less, we need to recognize what is distinctive about how we know now. And a good place to start is with this simple fact: the things we carry allow us to know more than ever about the world, faster than ever. But they also allow the world to know more about us—and in ways never dreamed of by Warren and Brandeis. Knowledge has become transparent. We look out the window of the Internet even as the Internet looks back in.

Most of the data being collected in the big data revolution is about us. "Cookies"—those insidious (and insidiously named) little Internet genies—have allowed websites to track our clicking for decades. Now much more sophisticated forms of data analysis allow the lords of big data, like Google and Amazon, to form detailed profiles of our preferences. That's what makes the now ubiquitous targeted ad possible. Searching for new shoes? Google knows—and will helpfully provide you with an ad showing a selection of the kind of shoes you are looking for the next time you visit nytimes.com. And you don't have to click to be tracked. The Internet of Things means that your smartphone is constantly spewing data that can be mined to find out how long you are in a store, which parts of the store you visit and for how long, and how much, on average, you spend and on what. Your new car's "black box" data recorder keeps track of how fast you are traveling, where you have traveled and whether you are wearing your seatbelt. That's on top of much older technologies that continue to see widespread use—such as the CCTV monitors that record events at millions of locations across the globe.

And, of course, data mining isn't done just for business purposes. Arguably, the United States' largest big data enterprise is run by the NSA, which was intercepting and storing an estimated 1.7 billion emails, phone calls and other types of communications *every single day* (and that was way back in 2010).[1] As I write this, the same organization is purported to be finishing the building of several huge research centers to store and analyze this data around the country, including staggeringly large million-square-foot facilities in remote areas of the United States.

We all understand that there is more known about what each of us thinks, feels and values than ever before. It can be hard to shake the feeling that we are living in an updated version of Jeremy Bentham's famous panopticon—an eighteenth-century building design that the philosopher suggested for a prison. The basic idea was a prison as a fishbowl. Observation, Bentham suggested, affects behavior—and prisoners would control their behavior more if they knew their privacy was completely gone, if they could be seen by and see everyone at all times.

In some ways, our digital lives are fishbowls; but fishbowls we've gotten into willingly. One of the more fascinating facts about the amount of tracking going on in the United States is that hardly anyone seems to care. That might be due not to underreporting or lack of Internet savvy by the public (although both are true) but to the simple fact that the vast majority of people are simply used to it. Moreover, there are lots of positives. Targeted ads can be helpful, and smartphones have become

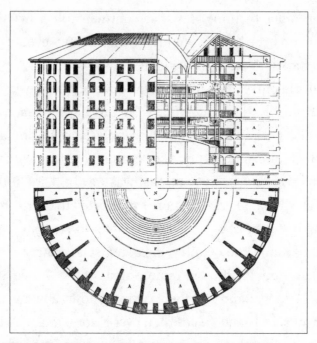

Fig. 2. Elevation, section and plan of Jeremy Bentham's
Panopticon penitentiary, drawn by Willey Reveley, 1791.

effectively indispensable for many of us. And few would deny
that increased security from terrorism is a good thing.

Partly for these reasons, writers like Jeremy Rifkin have been
saying that information privacy is a worn-out idea. In this view,
the Internet of Things exposes the value of privacy for what it is:
an idiosyncrasy of the industrial age.[2] So no wonder, the thought
goes, we are willing to trade it away—not only for security, but
for the increased freedom that comes with convenience.

This argument rings true because in some ways it *is* true: we
do, as a matter of fact, have more freedom because of the Internet and its box of wonders. But, as with many arguments that
support the status quo, one catches a whiff of desperate rational-

ization as well. In point of fact, there is a clear sense in which the increased transparency of our lives is not enhancing freedom but doing exactly the opposite—in ways that are often invisible.

The Values of Privacy

If you are arrested for a serious crime in the United States today, your picture is taken, you are fingerprinted, and in some precincts the inside of your cheek is swabbed in order to obtain a sample of your DNA. In his dissenting opinion in the recent Supreme Court case on DNA identification techniques, Supreme Court Justice Antonin Scalia argued that such techniques amount to illegal searches.[3] We are, he said, opening our mouths to government invasion and tyranny.

Legally speaking, the case was complicated by a lack of clarity over whether collecting DNA constitutes a search. That is partly because DNA collection does not require a cheek swab; it can be collected from the skin or hair, for example. Thus, as the majority opinion noted, it is unclear why fingerprints wouldn't also constitute an illegal search if DNA samples do.

What *is* clear is that photos, fingerprints and DNA samples allow the police to identify and reidentify you—in ways that are increasingly immune to deception or alteration. You can change your name and your appearance, and fingerprints are not actually unique. But you can't as easily mess with the DNA—it is, in a real sense, part of who you are.

Of course, being able to identify criminal suspects is generally a very good thing—and DNA has proven to be an effective tool not only in this regard, but also for exonerating innocent people of crimes for which they've been falsely accused (and sometimes convicted). At the same time, human beings have always been somewhat suspicious about new means of identification. That includes basic methods that we often overlook. Consider names. A rose may smell as sweet under any other name, but the fact that it has a name at all gives us an ability to reidentify it quickly and communicate that identification to one another. That's why, in some cultures, knowing someone's true name can give you (magical) power over them. You know how to identify "who they really are."

For similar reasons, images have often been said—as Warren and Brandeis were well aware—to have power. Even now, a photograph remains one of our best ways of identifying anything: we record in detail what our memories can't. Is it any wonder that individuals in some cultures were hesitant to allow their pictures to be taken? The idea that a camera could steal your spirit can be seen as a way of representing a real truth: that a picture identifies you, and like people's knowledge of your name is not something that is necessarily in your control.

This idea of control is closely connected to the idea of information privacy. The broad notion of privacy is difficult, if not impossible, to define in a straightforward way, and the narrower notion of information privacy is not much better. But even without a precise definition, it is clear that there are several marks or symptoms associated with information privacy. One of those concerns protection: we think of information as private to the extent that it is protected from interference or intrusion.[4] Another concerns control: information is private to the extent to which we control access to it.

94

Why do we value protecting and controlling our information? A cynic might say: we value it only when we have something to hide. But of course, even if this is true, it doesn't really answer the question. That's because it depends on *what* you are hiding and *whom* you are hiding it from. Hiding a criminal past is one thing; hiding Jews in your basement from the Nazis is another.

In reality, there are much more basic reasons information privacy matters to us.

The Pool of Information

In the summer of 2014, following the revelations of Edward Snowden, the *Washington Post* revealed what many had long suspected: that the NSA, in targeting foreign nationals, is collecting and storing extremely large amounts of information on American citizens.[5] This information is not restricted to metadata of the sort collected by the NSA's infamous phone data collection program. It is content—photos, Web chats, emails and the like.

U.S. law prevents the targeting of U.S. citizens without a warrant (even if it is just a warrant from the secret court established for this purpose by the Foreign Intelligence Surveillance Act (FISA) of 1978). But citizens' digital data is often vacuumed up "incidentally" when the NSA is collecting the posts, emails and so forth of legally designated foreign targets. Nothing currently prevents the NSA from engaging in this "incidental collection." And the incidentally collected data can be stored indefinitely. Moreover, no law prevents the agency—and other U.S. intelligence and law enforcement agencies—from accessing the incidentally collected content without a warrant, into perpetuity.

The storage of incidentally collected data seems clearly wrong. Yet the reasons that make it so also help to explain why as a nation we sometimes sympathize with the sentiment voiced by Representative Mike Rogers in 2013: that your privacy can't be violated if you don't know about it—a non sequitur of such numbing grossness that only Peeping Toms could have greeted it with anything other than laughter.

But before getting back to the NSA, let's do another thought experiment. Imagine for a moment that I could perform something like the Vulcan mind meld with you (okay, okay, I'm dating myself). I telepathically read all your conscious and unconscious thoughts and feelings. You don't share your thoughts with me; I take them. I'm sure you'd agree that such an act of mental invasion would be wrong and harmful, but let's think about why.

The first and most obvious reason is that it has potentially dangerous consequences for you. And, Mike Rogers aside, that danger exists whether or not you know about my violation. Suppose you don't know. The more I know about you and the less you know about my knowledge, the easier it could be for me to take advantage of your ignorance: and the easier you will be to control or exploit.

Intentions matter. The fact that I can read your thoughts doesn't necessarily mean I will exploit you. I may be purely motivated by science: I might jot down your thoughts and do nothing to profit from them. Or I may be like Professor X of the comic book heroes X-Men, only motivated by truth, justice and the American way. More simply, the fact that I know what you like may help me guide you toward experiences you haven't had yet but would enjoy. Think of Amazon: part of their business model is predicated on acquiring information about their customers' preferences—information often obtained without the customer really knowing it. They use this information to predict what else the customer might like—and to ensure that the customer is given every opportunity to buy it. There is no doubt that many of us feel uncomfortable about the amount of information collected now by corporations for the purpose of selling us stuff we need (or making us want to buy stuff others would like us to need). But it doesn't necessarily involve nefarious motives. So intentions matter, and even in the case of mind-reading, it is not absolutely certain that bad things will happen, even if you don't know about it.

But what if you *do* know that I am reading your thoughts? Well, you'll be wary, naturally. So wary that you will likely try and censor your thoughts and even your activities—perhaps by humming some Mozart in your mind to disguise your thoughts as best you can. And the reason you'd do so would be obvious— no matter how good I may seem to you now, you will want to minimize your exposure to exploitation and manipulation. This is not surprising. As Bentham knew when he designed his panopticon, observation affects behavior. But, of course, that too isn't necessarily bad. It is why security cameras are not always hidden. If you know you are being watched, you are less likely to act out. And that can be an instrument for good. Or not.

So, one reason privacy is important is that invasions of it can lead to exploitation, manipulation and loss of liberty. These, in turn, obviously can negatively affect a person's autonomy. But the possibility of bad effects is one thing, the actuality another. This is precisely the point that defenders of the NSA programs, for example, have been at pains to make. For all that's been said so far, there *might* be negative consequences from, e.g., the NSA's policy of massive incidental collection and other data-sweep programs, *if* the agency or its architects were assumed to have bad or corrupt intentions. But why, some say, should we think that?

The fact is, however, that we don't have to know anything about the intentions of the program's architects in order to be worried. The NSA programs are dangerous to democracy even if we assume that their architects were motivated by the best of intentions—as no doubt many of them were. Roads to unpleasant places are frequently paved with the sweetest of intentions.

The NSA database could be described as a pool of information. This is an apt metaphor. In law, swimming pools are called attractive nuisances. They attract children and, as a result, if you own a pool, even if you are a watchful, responsible parent yourself, you still have to put up a fence. Similarly, even if we can trust that the architects of the NSA's various programs had no intention of abusing the information they are collecting about American citizens, the pool of information could easily prove irresistible. And the bigger the pool, the more irresistible it is likely to become. This is not just common sense, it explains why the NSA's repeated assertions that they aren't actually looking at the content of emails, or targeting Americans, should have been greeted with skepticism. The pool of data is a pool of knowledge. Knowledge is power; and power corrupts. It is difficult to avoid drawing the inference that *absolute knowledge might corrupt absolutely.*

That, not surprisingly, is the view of folks like Edward Snowden. But a growing number of stories strongly suggest that fear of abuse is more than a mere theoretical worry. These examples are not constrained to the widely reported cases of NSA employees using their access to spy on sexual partners,[6] nor to similar cases in the UK where analysts collected sexually explicit photos of citizens without cause. More troubling, if less titillating, is the fact that the secret FISA court itself has complained that the NSA misrepresented its compliance with the court's previous rulings that various NSA techniques were unconstitutional.[7] In other words, the FISA court is being ignored by the very agency it is assigned to oversee and monitor. It is hard not to form the impression of an agency that feels it knows better than the judiciary or the Congress. And that, surely, should be worrying.

But the most disturbing fact is the massive continued storage of incidentally collected content itself (again: emails, photos, chat conversations and so on)—information that, as reported in the *Post*, is routinely searched by the CIA and the FBI—all without a warrant, even from the ineffective FISA court, and without any real oversight. Such searches needn't even be reported, and there is, presently, no legal oversight to prevent queries that are unrelated to national security, or even motivated by political ends. And relying on the agencies themselves to report abuses is like relying on the tobacco companies to tell us whether smoking is harmful.

While that's one of the major problems with the NSA collecting massive amounts of incidental information about Americans, it also helps explain why people don't seem too concerned.

Putting up fences is arduous, time-consuming and expensive. And it does cut down on easy access to the water. So, if you want to get in that pool with the best intentions—you want to find the terrorists—it is natural to think that the fence only gets in the way of what matters. If you trust that is what the owners of the pool are after, then worries about possible long-term negative consequences will seem abstract and, well, philosophical. After all, it is pretty clear human beings find it difficult to think about long-term consequences—that's true whether we are talking about swimming pools or global warming. If nothing bad has happened already that we know about as a result of privacy invasions, then what's the problem?

Unfortunately, if the pool of information about American citizens is systematically abused we aren't going to know about it—at least, not easily. When it comes to global warming, at least we'll get to realize the consequences of our current policies (or lack of them) one way or another. But the abuse of knowledge isn't going to be so obvious, and the abusers will have every reason to hide behind good intentions. That was one of the points made by the President's own review panel's report in 2013.[8] That panel—made up of not only writers and scholars including Cass Sunstein but former leaders of the CIA—suggested, in fact, more than simply fencing the pool (passing legislation to make it more difficult to access); they suggested the pool be drained. That is, they urged that all incidentally collected information (again, mostly on Americans, and far outweighing the amount being collected on warranted targets) simply be removed from the NSA's databases. This has not yet been done.[9]

Privacy and the Concept of a Person

The potential dangers of abusing big data are one reason the storage of incidentally collected information is wrong. But there is another: the more insidious harm is not "instrumental" but "in principle."

Just this point was made over half a century ago in one of the most cited discussions of the right of privacy. In 1965, Edward J. Bloustein argued in a paper that what is wrong with such intrusions is

> not the intentional infliction of mental distress but rather a blow to human dignity, an assault on human personality. Eavesdropping and wiretapping, entry into another's home, may be the occasion and cause of distress and embarrassment but it is not what makes these acts of intrusion wrongful. They are wrongful because they are demeaning of individuality and they are such whether or not they cause emotional trauma.[10]

Following what he took as the main point of Warren and Brandeis, Bloustein grounded the right of informational privacy on the intrinsic value of human individuality. The connection was what he called "personal freedom":

> The fundamental fact is that our Western culture defines individuality as including the right to be free from certain types of intrusions. This measure of personal isolation and personal control over the doctrines of its abandonment is

of the very essence of personal freedom and dignity, is part of what our culture means by these concepts.[11]

Let's unpack this thought. Philosophers have traditionally distinguished freedom of choice or action from what we might call the autonomy of decision. To see the difference, think about impulse buying. You may "freely" click on the Buy button in the heat of the moment—indeed, corporations count on it—without that decision reflecting what really matters to you in the long run. Decisions like that might be "free" but they are not fully autonomous. Someone who makes a fully autonomous decision, in contrast, is committed to that decision; she owns it. Were she to reflect on the matter, she would endorse the decision as reflecting her deepest values.

Totally autonomous decisions are no doubt extremely rare; indeed, philosophers have long questioned whether they are possible at all. But it is clear that we value autonomy of decision, even if we can only approximate the ideal. That's because autonomy of decision is part of what it is to be a fully mature person. And that, I believe, tells us something about why privacy matters. It matters, at least in part, because information privacy is linked to autonomy, and thereby an important feature of personhood itself.

There are two ways to infringe on a person's autonomy of decision. The most obvious way is by *overruling* the decision, either by direct compulsion (I point a gun at your head) or by indirectly controlling your values and commitments (I brainwash you). A subtler way of infringing on your autonomy is to *undermine* it. Suppose a doctor makes the decision to give you a drug without asking your permission. Nobody has made you decide to do something. But your autonomy has been undermined nonetheless, and for an obvious reason: your decision has been foreclosed. You are not in a position to make the decision. It has been made for you.

Apply this to privacy. One mark of information privacy is control; that is, you have at least some *control over how and to whom you share those aspects of your self.* So consider a limit case. If you have a condition that compels you to say out loud every thought that comes into your head—whether you like it or not—your autonomy of decision has been overruled. You are at the mercy of your condition.

But privacy invasions generally don't harm autonomy in this way. They don't overrule privacy. They undermine it. Suppose, to take a more old-fashioned example, that I break into your house and read your diary over and over again, every day. Suppose further that I make copies for my friends. Even if, again, you never learn of this, I am harming you in a new way—by undermining your capacity to control your private information. Whether you know it or not, that capacity is diminished. You may *think* you have the autonomy to decide whether to share your diary or not. But in fact, you are not in a *position* to make the decision; I've made that decision for you. Your autonomy of decision has been undermined.

As noted above, part of what makes your individual mind *your* mind is that you have a degree of privileged access to your mental states. And that includes, crucially, the ability to control access to the content of those thoughts and feelings—to choose whether and when you share this information with others. Part of the reason we value having this control is because it is a necessary condition for being in a position to make autonomous decisions, for our ability to determine who and what we are as persons.

It is here we see the danger inherent in systematic and sweeping collection by the government of the private information of citizens who have neither been charged with nor suspected of a crime. Such intrusions on information privacy—whether or not that information is acted upon or whether the intrusion is known—undermine not only dignity but, depending on how systematic the intrusions turn out to be, your actual capacity to control how and what information you share with others. The harm to your autonomy of decision becomes more global.

The systemic nature of the invasion therefore matters. Again, this point can be made starkly by looking back at our telepathic case, where I read your mind without your consent. Suppose it happens only once. Obviously, if I act on this knowledge in order to manipulate and control you, then I may directly harm your autonomy. And the risk of this happening may be great enough to prevent me from being tempted to use my power. Moreover, as I've been emphasizing, it shows a lack of respect for your status as a person, an autonomous being. But if I only read your mind—or your diary—once (and do nothing with the data), then presumably your autonomy itself is not affected. You don't become less in control of your self or face a diminished capacity of any sort.

Now let's suppose this happens again and again over time, in an organized way. From my perspective—the perspective of the knower—your existence as a distinct person will begin to shrink.

Our relationship will be so lopsided that I may well cease to regard you as a full subject, as a master of your own destiny. As I learn what reactions you have to stimuli, why you do what you do, you may become like any other object to be manipulated *even if I do not, in fact, manipulate you.* You may be, as we say, dehumanized in my eyes. The connection between a loss of privacy and dehumanization is a well-known and ancient fact, and one which we don't need to appeal to science fiction to illustrate. It is employed the world over in every prison and detention camp. It is at the root of interrogation techniques that begin by stripping people literally and figuratively of everything they own.

The connection between autonomy and privacy may sound surprising to some. After all, one could say: we are in fact willing to trade away our privacy as never before, precisely for the purpose of *increasing* autonomy. Our willingness—the thought goes—to trade privacy for security is just one example of this phenomenon. Another is our near total passivity when it comes to the trading of our data for profit by private corporations. We want more autonomy and they are providing it, by giving us convenience. Indeed, that's precisely the business model of corporations like Facebook or Amazon—to maximize convenience and anticipate our needs. Thus, one might say, it is not surprising that we click past all the privacy policies on the Web because we want the choices, the convenience—the autonomy—that only the playground of the infosphere can bring. Privacy suffers, but autonomy increases.

This argument, however, gets things the wrong way around. When we systematically collect private data about someone, we implicitly adopt what the philosopher Peter Strawson called the "objective" or detached attitude toward her.[12] We see her as something to be manipulated or controlled—even if, in fact, we never get around to the actual manipulating or controlling. Where privacy is limited in the detention camp or prison, the adoption of this attitude toward the inmate is of course explicit. It is an intrinsic feature of the enterprise and it is intuitively felt as such by those detained. Crucially, however, it remains implicit in more subtle invasions of privacy. In some cases, this is unsurprising. When a business sells or otherwise profits from your private information—your Web searches, for example, or email address—it intentionally treats you as an object: an object of profit. Indeed, the nominal idea behind the privacy policies none of us read is to inform us of how our information will be used. They are a nod to our status as autonomous beings.

In truth, however, the Internet of Us is making privacy policies moot. When almost every object we interact with is wired, it becomes useless to assume that we consent to the mining of the data trail attached to our use of that object. That's because we simply have no way of being able to anticipate how the data being extracted from our refrigerators, for example, might be used in the future—by a company or by a government. Once the data is out there, it is out there. Any illusion we might have had about controlling or owning it gradually disappears. As Sue Halpern, an astute observer of the digital age, remarks: "The Internet of Things creates the perfect conditions to bolster and expand the surveillance state. In the world of the Internet of Things, your car, your heating system, your refrigerator, your fitness apps, your credit card, your television set, your window shades, your scale, your medications, your camera, your heart rate monitor, your electric toothbrush, and your washing machine—to say nothing of your phone—generate a continuous stream of data that resides largely out of reach of the individual but not of those willing to pay for it or in other ways commandeer it."[13]

Earlier I noted there are two marks to information privacy: control and protection. Control over our information may be increasingly under threat by the Internet of Things. But that only makes concentrating on restricting and regulating information flow all the more important. The Internet of Things is enlarging the pool of data and information available for future use; that's why we need more fencing. We need the fences of regulations not only because they help prevent abuses, but because the pool threatens our autonomy.

There is another point here as well. Surveillance treats us as means, not as ends. And that is another reason the incidental collection of our data should worry us. A government that sees its citizens' private information as subject to tracking and collection has implicitly adopted a stance toward those citizens inconsistent with the respect due to their inherent dignity as autonomous individuals. It has begun to see them not as persons but as objects to be understood and controlled. That attitude is inconsistent with the demands of democracy itself.

Transparency and Power

Invasions of privacy aren't always wrong. If they were, we wouldn't have to spend so much time talking about the issue. My point is that they are always *pro tanto* wrong, as the legal scholars say. They are wrong—but wrong other things being equal.

Invasions of privacy can therefore be justified in the overall context. Searches of people's homes are judged "warranted" (that is, justified) for all sorts of reasons by the courts, as are surveillance operations of criminal suspects. Or consider the case of metal detectors and full body scanners at airports. The latter were (and still are) controversial on privacy grounds; moreover, more than one person argued that the scanner violated their dignity. But while scans like this can make you uncomfortable, this sort of directed, publicly known invasion of one's privacy is not equivalent to the systematic program of incidental collection and meta-analysis of phone call data practiced by the NSA. That's because full body scans are given to commercial airplane passengers for a very specific reason: to detect whether they have a concealed weapon or explosives. This reason is well understood—or should be—by those given the scans. It is, in fact, a classic case of trading privacy for more security. It is a trade that may be justified, all things considered. Airport body scans are not stored indefinitely and open to the scrutiny of security agencies. They are made, examined, and eliminated. And they aren't being done secretly either. A better analogy would be this: secret scanners are set up so scans are taken of every person in his or her home. No one is told about the scans. They are stored indefinitely, and a wide range of agencies can examine them without a warrant. Still think that would be justified?

The possible negative consequences of losses of privacy in the digital age suggest that we must prepare for the worst even as we hope for the best. Think again of the swimming pool example.

We need fences around our digital pools of information too. That's why, for example, some of the steps recently proposed by the Obama administration—to strengthen the FISA court's powers, and to limit some of the NSA's surveillance programs—are at least steps in the right direction.[14]

No one denies that governments naturally diminish our autonomy in all sorts of ways. Just participating in a government, as Hobbes stressed, is a trade-off. But the point I've been making in this chapter is that there is something different in the case of *systematic, unknown* invasions of privacy. By invading our privacy without our knowledge, governments are making invisible decisions for the citizenry as a whole. That's not the same as restricting autonomy by asking people to go through a scanner at the airport. That's power visible to all, applied to all. Nor is it like wiretapping a particular citizen whom the courts have decided is a potential danger. Rather, these systematic, unknown invasions of privacy treat the citizenry as a whole in an unhealthy way. We are being regarded as unworthy of making up our own minds, whether we know it or not. That is an attitude that is corrosive of democracy, one made all the more corrosive by not being visible.

These reflections also give the lie to the idea that privacy of information is a modern creation. It is not. The source of privacy's value is deeper, lying at the intersection of autonomy and personhood itself. That is why privacy still matters. We are wise not to forget that, even as we trade it away.

Knowledge may be transparent, but power rarely is.

6.

Who Does Know: Crowds, Clouds and Networks

Dead Metaphors

Truths, Nietzsche once wrote, are worn-out metaphors, "coins that have lost their pictures, and now only matter as metal, not as coins."[1] The word "network" has lost its luster in just this way: we now just accept it as a literal description of the facts. Our economy is a network; our social relations are networked; our brains are composed of neural networks; and of course, the Internet, the World Wide Web, is a network. Thus, we might wonder whether knowledge is too. This idea has become reasonably common in tech circles. Some believe it is a game-changer. Again, David Weinberger is at the forefront: "In a networked world, knowledge lives not in books or in heads but in the network itself."[2] Indeed, in Weinberger's view, the information age is basically over. We live in the networked age, where information doesn't come in discrete packets but in structured wholes.

Let's start unpacking that notion by looking at the idea of a network itself. Think of the ways in which one can describe—or map—a transportation system, such as a subway. One way is to simply superimpose the path of the train tracks onto an existing street map. That works fine, as long as the street map is not too detailed itself, and as long as there aren't too many underground tubes and tracks. If, for example, there is just one track, with two stops, then passengers only need to know where these stops are in order to orient themselves. But what if there are dozens of stops, and the lines crisscross and don't follow the paths of the streets overhead? That was the problem that Harry Beck, an employee of the London Underground, aimed to solve in 1931 by developing a new Tube map—one which, with additions, is still familiar to riders today. What was different about Beck's map is that he ignored the geography of the city and concentrated solely on showing, without reference to scale, the sequence of stations and the intersection of the Underground lines.

By doing so, Beck was able to bring to the fore the information that Tube riders really wanted most: how many stops are in between the present stop and the one you want to get to, and where the lines interconnect. By knowing these two facts, you can deduce how to get from A to B.

As the information theorists Guido Caldarelli and Michele Catanzaro note, Beck's map is like a graph. As such, it displays a basic feature of a network: "in networks, topology is more important than metrics. That is, what is connected to what is more important than how far apart two things are: in other words, the physical geography is less important then the 'netography' of the graph."[3] The reason why, in this case, is pretty clear. The netography or topology of the Underground matters to us because what we are interested in is how information is distributed in that system—or, more bluntly, in how we riders are distributed along the lines of the Underground tracks. What Beck's map shows is that thinking of something as a network is useful when what matters is a complex pattern of distribution between points rather than the points (the "nodes") themselves. This is part of the reason it makes sense to say that knowledge is becoming more and more networked. The infosphere has made it possible to distribute information so efficiently, and so quickly, that these facts about the distribution become important in themselves.

But really, we are more networked than that. We are increasingly *composing* a knowledge network—or is it composing us?

Knowledge Ain't Just in (Your) Head

Let's go back to neuromedia. What would happen if it became available to the general population? The nature of communication would change, certainly. But that's not all. The boundaries between ourselves and others would, in certain key respects, change as well—especially with regard to how we come to know about the world.

Suppose everyone in a particular community has access to this technology. They can query Google and its riches "internally"; they can comment on one another's blog posts using "internal" commands. In short, they can share knowledge—

they can review one another's testimony—in a purely internal fashion. This would have, to put it lightly, an explosive effect on each individual's "body of knowledge." That's because whatever I "post" mentally would then be mentally and almost instantly accessible by you (in a way that would be, we might imagine, similar to accessing memory). We'd share a body of knowledge by virtue of being part of a network. *But that is not the most drastic fallout of neuromedia.* The more radical thought is that we are sharing the very cognitive processes that allow us to form our opinions. And to the extent that those processes are trustworthy and accurate, we can say we are sharing ways of knowing.

The traditional view has always been that humans know via processes such as vision, hearing, memory and so on. These ways of getting information are internal; they are in the head, so to speak. But if you had neuromedia, the division between ways of forming beliefs that are internal and ways that are not would no longer be clear. The process by which you access posts on a webpage would be as internal as access to your own memory. So, plausibly, if you come to know, or even justifiably believe, something based on information you've downloaded via neuromedia, that's not just a matter of what is happening in your own head. It will depend on whether the source you are downloading from is reliable—and that source will include the neural networks and cognitive processes of other people. In short, were we to have neuromedia, the difference between relying on yourself for knowledge and relying on others for knowledge would be a difference that would make less of a difference.

Andy Clark and David Chalmers' "extended mind" hypothesis suggests that, in fact, our minds are *already* extended past the boundaries of our skin.[4] When we remember what we are looking for in a store by consulting a shopping list on our phone, they argue, our mental state of remembering to buy bread is spread out; part of that state is neural, and part of it is digital. The phone's notes app is part of my remembering. If Clark and Chalmers are right, then neuromedia doesn't extend the mind any more than it already is extended. We already share minds when I consult your memory and you consult mine.

The extended mind hypothesis is undoubtedly interesting, and it may just be true. But we don't actually have to go so far to think knowledge is extended. Even if we don't literally share minds (now, at least), we do share the processes that ground or justify what our individual minds believe and think. As philosopher Sandy Goldberg has pointed out, when I come to believe something based on information you've given me, whether or not I'm justified in that belief doesn't depend just on what is going on in *my* brain. Part of what justifies my belief is whether *you*, the teacher, are a reliable source. What justifies my receptive beliefs on the relevant topic—what grounds them—is the reliability of a process that includes the teacher's expertise. So whether *I* know something in the receptive sense already can depend as much on what is going on with the teacher as it does the student.[5]

Goldberg's hypothesis seems particularly apt when we form beliefs receptively via digital sources—which, as I said, can be understood as knowing via testimony. In relying on

TripAdvisor, or Google Maps, or Reddit, I form beliefs by a process that is essentially socially embedded—a process the elements of which include not just chips and bits but aspects of other people's minds, social norms and my own cognition and visual cortex. How I know is already entangled with how you know.

The Knowing Crowd

So far then, we've seen that knowledge has become increasingly networked in at least two discernible ways: Google-knowing is the result of a network. And our cognitive processes are increasingly entangled with those of other people.

This raises an obvious question. Is it possible that the smartest guy in the room *is* the room? That is, can networks themselves know?

There are a few different ways to approach this question. One way has to do with what those in the AI (artificial intelligence) biz call "the singularity"—a term usually credited to the mathematician John von Neumann. The basic idea is that at some point machines—particularly computer networks—will become intelligent enough to become self-aware, and powerful enough to take control.

The possibility of the singularity raises a host of interesting philosophical questions, but I want to focus on one issue that is already with us. As we've discussed, there are reasons to think that we digital humans are, in a very real sense, components of a network already. So, could networked groups literally know things over and above what their individual members know?

And if groups know things as a collective—in any sense of "know"—then they have to be able to have their own true, justified beliefs. Is that possible?

Some philosophers have argued that it is, and cite the fact that groups can pass judgments even when no individual in the group agrees with the judgment. For example, imagine a group of interviewers trying to choose the best person for the job. Suppose they interview three candidates and each of the interviewers ranks each of the candidates by order (with one being highest). It might turn out that nobody ranks candidate B as number one but that B still turns out as the candidate with the highest *cumulative* ranking (if, for example, everyone ranks B second but split their remaining votes). If so, then the group "believes" that B is the best candidate for the job even though no individual in the group has ranked that candidate number one.

The eminent philosopher of sociology Margaret Gilbert has argued that, if they exist, real group beliefs are the product of what she calls "joint commitments."[6] A joint commitment is the result of two or more people expressing a readiness to do something together as a unit—like dancing a waltz, performing a play, starting a business, or interviewing a job applicant. You don't, Gilbert emphasizes, always have to engage in a joint commitment deliberately. Often we express our willingness to act together only implicitly, as I might if I just held out my hand to you and gestured toward the dance floor. But however individuals express their readiness to jointly commit, their expression must be common knowledge to all; it must be something that is so taken for granted that everyone knows and everyone knows that everyone knows. In Gilbert's view, when these conditions are in place and a group has a joint commitment of this sort, it makes sense to think of groups as having a belief just as individuals have beliefs.

Gilbert's hypothesis explains why we do sometimes hold groups responsible over and above their members. As I write this, the corporation British Petroleum received a billion-dollar-plus fine for its role in the Deepwater Horizon oil spill in the Gulf of Mexico. Corporations, while they might be treated as "legal" people, are actually groups of people jointly committed to a common end of profiting from a particular enterprise or enterprises. When we hold groups who are jointly committed in this way responsible, we are holding a group responsible, not the individuals within the group. And it does seem as if we hold groups responsible not just for their actions but for their views— for example, if our job interviewers were, as a group, to believe that a man was the best applicant for the job even though a woman with far better credentials had applied. In such a case, we might hold that the belief—no matter how sincere—was unreasonable.

In some cases, Gilbert's view of joint commitments may also explain some group commitments made by digital humans. The digital "groups" that we form are often bolstered by a joint commitment to something, whether it be a political ideology, a hobby, a sport, or the practice and theory of hate. Such groups often do have the sort of common knowledge that joint commitment requires. But it is less clear whether people participating in Internet chat rooms, or posting on a comment thread on a popular blog, are really intending to "do something together." Sometimes that may indeed be the case—Wikipedia is a good example of a network where posters are committed to a joint enterprise—but often the opposite is true. From the standpoint of Gilbert's theory, Internet groups and networks may not have any group knowledge at all.[7]

Yet even if social networks don't literally know like individuals do—a view that Weinberger himself shies away from— there is still another way of thinking about the question of whether networks know. Groups can certainly *generate* knowledge, in the sense that the aggregating of individual opinions can give us information, and possibly accurate, reliable information, that no one individual could. Consider that ubiquitous feature of your online life: the ranking. There was a day when the only way to get information on whether a movie, restaurant or book was to your taste was to consult a professional review. Now we also have the star system. Instead of one review, we can get dozens, hundreds or even thousands. And in addition to the "qualitative" comments, we get an overall ranking, the average of individual rankings assigned to the product. Useful? Certainly. And most of us know some simple facts about such systems as well. To name the most obvious: the more rankings, the more reliable we tend to take the average score to be (1,000 rankings with an average of 4 stars is far more impressive than three rankings of 4.5 stars). Of course, we also know that the fact that many people like something doesn't mean we'll like it too.

The fact that we so often trust such rankings—at least, in the right conditions—points out that we already tend to abide by the main lesson of James Surowiecki's 2004 landmark book *The Wisdom of Crowds*. Surowiecki's point was that in certain conditions, the aggregated answers of large groups could be wiser—could display more knowledge—than an individual, even an individual expert. Suroweicki's most famous example comes from the work of Francis Galton, a British scientist. Galton examined a competition in which 787 contestants at a country fair estimated the weight of an ox. The average of all guesses was 1,197 pounds. The ox weighed 1,198 pounds.[8]

Another of the most famous results in social science helps explain the limits and lessons we can draw from examples like the weight of the ox. Suppose a group of people vote on a yes-or-no question, where only one of the answers can be right. Suppose too that the probability that any one person gets the answer right is over 50%. According to what is called the Condorcet Jury Theorem, the larger the group, the more the probability of a correct answer by a majority of the group goes upward or approaches 100%. The basic math is intuitive: as the probability of a right answer by an individual goes upward, the probability that the collective answer is correct also rises (where the correct answer is decided by majority vote). So, if you have enough people, then even if they are only a little better than chance at getting it right, the group can be exceedingly good at tracking the right answer.

There's a hitch, however. Groups do better than individuals only under certain conditions, including the assumptions we stated at the outset: each individual is better than chance at getting the right answer, and the answers are aggregated by majority rule. In addition, the theorem applies best when the individuals in question (the "voters") are independent of one another, and not (at least in a statistically meaningful way) influenced by other voters' decisions (thus lowering the chances that they are participating in information cascades).

There is considerable debate about the extent to which the Condorcet Jury Theorem maps onto real-life situations. One question, for example, is whether it can help justify the thought that democratic institutions are reliable mechanisms, other things being equal, for determining the best public policy. The theorem gives some comfort to that idea—again, assuming that the conditions are met. And sometimes they are. Voters in many elections *are* unaware of the votes of others at the time they cast their vote. (Although, famously, turnout on the West Coast of the United States can be dampened in presidential elections due to the news media's polling of voters in later time zones. And polling results in general may shift voting patterns.) Relatedly, in some online rankings, consumers may be reasonably independent in their decision-making.

Of course, it is just as often the case that these assumptions are not met. First, people are often *not* better than chance at judging the truth. We are all susceptible to bias and prejudice, and our opinions are often not all that independent (causally or statistically). Second, even *if* the members of a particular crowd are a bit better than chance at judging correctly, that doesn't help much if the "crowd" in question is pretty small. And third, sometimes we might even be worse than chance. And in those situations where people are worse than chance at judging a situation, then the more people you get to answer the question, the higher the probability that you'll get the *wrong* answer. In that case, the crowd is not wise but unwise.

This brings us to a key point: whether or not a network "knows" something (even in the nonliteral sense) depends on the cognitive capacities (and incapacities) of the nodes on that network—the individual people who make it up.

A good example is prediction markets. Markets like this trade in futures, but participants aren't betting on whether a given company's monetary value will rise but whether, for example, a politician will win an election or a particular movie will win an Oscar. These markets had some notable early successes; Intrade, for example, was famously better at predicting the 2006 midterm elections than cable news. (Intrade was one of the most widely cited before it closed in 2013.) In a certain obvious sense, markets like this can be seen as encoding the information of the network of investors that make it up—not only about what may happen in the future but about the state of an election at any given time. But the way in which this information is aggregated is not, as in the cases above, statistical. Prediction markets don't average the views of their participants, rather they work in the same way other markets work: the more confident buyers are that a given candidate will win, the higher his or her "stock" goes— the more value is attached to it, no matter how many people may own that stock.

But prediction markets also have their limits. A well-known example, noted by David Leonhardt of the *New York Times*, was the 2012 Supreme Court decision on the Affordable Care

Act.[9] Right up to the last minute, Intrade was indicating a 75 percent chance that the Act's mandate would be declared unconstitutional. That was wrong—and, in fact, as Leonhardt noted, many insiders had been going the other way. Arguably the insiders' information was better, and their take on it more legally sophisticated, than that of the larger crowd. In this sort of case, the larger crowd is not the one you want to listen to. One might think the same goes for predicting something like the success of medical surgery. Unless the crowd has the same information and training as the relevant experts, it is not clear that they have wisdom to impart. As Leonhardt's colleague Nate Silver noted during the final run-up to the 2012 election, such markets may contain more or less sophisticated participants, and the more sophisticated the average participant, the more other sophisticated participants tend to trust it. Moreover, when a given market is highly cited in the press, "that opens up the possibility that someone could place a wager on [a candidate] in order to influence the news media's perceptions about which candidate has the momentum."[10] If so, the market may not be reflecting or mapping voter opinion but helping to determine it.

So, although networks can embody knowledge, or at least true information, not held by any particular individual, the extent to which they do so depends very much on the cognitive capacities of the individuals that make them up. *You can't take the individual out of the equation.*

The importance of the individual remains even though what we know as individuals depends on the social networks to which we belong. Take two indistinguishable people, Alycia and Bri, with the same belief and all the same evidence available by intro-spection. Stipulate that they are equally good (or bad) bullshit spotters, equally good (or bad) detectors of reliable testimony.[11] Suppose each is hooked in to a different online community: different friends on Facebook, different Twitter feeds, different news stations and so on. If Alycia's social network has high standards for belief and Bri's network has very lax standards, then there will be more unreliable testimonies floating around in Bri's network than Alycia's. That's because Alycia lives in a network where people in general are more critical and discerning. So more folks will just believe less, period. And what that means is that since Alycia will be getting at least some information from her network, a higher *percentage* of her information will be accurate even if she may have fewer firm opinions and beliefs overall. The opposite is true of Bri's community. They are more inclined to believe what others say just because they say it. As a result, Bri's social network will have more opinions—they *might*, depending on how lucky they are, even have more true opinions. But as a result, it is likely that a lower percentage of Bri's total number of beliefs will be true. So Bri's beliefs that are formed on the basis of testimony are less safe than Alycia's; they are more easily wrong, and more prone to be right by luck when they are right at all.

How much that matters will depend, of course, on what's at issue. When the question is which cats are cuter than other cats, we can afford to shrug our shoulders. *But when the stakes are high—when the questions concern matters like whether climate change is real or whether the measles vaccine causes autism—the situation is different.* Our community, our network, is only as smart as its standards for evidence allow. Even if you are as tough-minded as they come, if your social network is gullible, then you are more likely to receive unsafe testimony—and thus you are less likely to know. And that leaves us with a very clear lesson: our standards and epistemic principles matter. *Reasonableness* matters; being critical matters. And that, in turn, shows that what's important isn't just how we "train" our networks, but how we train the individuals that compose them. It is in our joint interest to support institutions that encourage the pursuit of critical public discourse by individuals.

In other, blunter words: the growing networked nature of knowledge makes the independent thinker more, not less, important than ever before. We need more of them.

The "Netography" of Knowledge

So far, we've left out a subtler but possibly more important point: our networked lives might be altering the very structure of knowledge.

Weinberger puts the matter this way:

Our system of knowledge is a clever adaption to the fact that our environment is too big to be known by any one person. A species that gets answers and can then stop asking is able to free itself for new inquiries. . . . [T]his strategy is perfectly adapted to paper-based knowledge. Books are designed to contain all the information required to stop inquiries within the book's topic. . . . [But with our new] connective medium . . . our strategy is changing. And it is changing the very shape of knowledge.[12]

Weinberger's point is that we've traditionally seen building or expanding the body of knowledge as the expansion of a series of "stopping points." Inquiry, scientific or otherwise, was aimed at getting to someplace safe, an answer we could trust. Once we got there, we could move on. But in Weinberger's view, the new mediums for knowledge created by digital technology are changing this picture. That's because, in his view, the Internet doesn't really deal with stopping points.

The notion that knowledge can have a shape—a structure—goes back a long way in Western culture. Plato himself drew something of a graph of knowledge, his so-called "divided line," which depicted a difference between true knowledge and mere opinion. True knowledge was founded, Plato thought, on a grasp of the eternal essences he called the Forms—another structural metaphor. Ultimate reality was comprised of the Forms, and hence the only true knowledge started with knowledge of them.

Since the seventeenth century, the dominant structural metaphor for knowledge has been architectural. Again, René Descartes deserves much of the credit (or the blame, depending).[13] According to so-called Cartesian foundationalism, the structure of knowledge is like a building or pyramid, with the foundation supporting the upper floors. Similarly, our beliefs are supported by other beliefs and ultimately by foundational beliefs and principles. Descartes' view was that if our house of belief was going to be stable and lasting, it had to end at certain propositions that were so obviously true that they were beyond any shadow of a doubt. These were the foundation stones—the ultimate stopping points.

The classic problem with Descartes' own version of this network was that he was a picky mason. Foundation stones that met his high standards were hard to find. To his mind, a foundational belief had to be certain and self-evident. His most famous example of the perfect foundation stone was a belief about himself: *I think, I exist*, is necessarily true whenever any of us thinks it. And that seems right: if anything is self-evident, it is that I think, for the simple reason that I can't doubt that I think without thinking. It is, as it were, a thought I can't escape. The problem is that this doesn't get us very far.

Later philosophers have tended to be less picky, but many stuck with the metaphor. Philosophers like Locke sensibly emphasized that the foundational nodes also had to include experience with an objective world. What grounded our beliefs, ultimately, was logic *and* experience. This is an idea that has sunk deep into our cultural bones: beliefs are justified, we think, when they are "supported" or "grounded." These words themselves reflect the foundationalist perspective of Descartes.

If you think of the body of knowledge as having a foundational structure, you'll be apt to be careful about what you add to the body of your beliefs. You'll want to make sure that new additions are secure, safe and overwhelmingly likely to be true. Otherwise they may upset the stability of the structure as a whole. As Weinberger argues, it is a way of thinking about knowledge that would come naturally in a world where *knowledge is expensive*—where recording and storing knowledge is itself a costly project. As he puts it, "Traditional knowledge has been an accident of paper." When the results of what we know can only be recorded slowly, when data must be written down, then the cost of that recording raises standards as to what we collect. We might want to collect all that is worth knowing, but libraries are finite physical spaces. They cost money to maintain. And so they require gatekeepers and filters to decide what gets into the library. And that encourages a certain picture of how knowledge fits together.

But in the infosphere, things look different. First, the library of the Internet is vast. The body of information available to us is so bloated that it would totter on *any* foundation, no matter how strong. Walls can't contain the digital library that surrounds us. And, of course, that information is growing every second.

Moreover, there is very little in the way of "gatekeeping" on the Internet. When ISIS beheaded an American journalist in 2014, displaying it on the Internet, digital giant YouTube responded by dropping access to the video and Google blocked searches of it. But that didn't stop it from continuing to get out. In the digital realm, information, even bad, morally reprehensible information, always finds a way.

Add these points to the facts we've discussed above—that not only what we know but also how we know is networked—and one can sympathize with the thought that it is no longer accurate, or even useful, to think of knowledge and justification as having a pyramid structure. Perhaps knowledge has no foundations. Indeed, Weinberger goes so far as to suggest that knowledge no longer rests on facts of any kind: "the idea that the house of knowledge is built on foundations of facts is not itself a fact. It is an idea with a history that is now taking a sharp turn."[14]

Actually, suspicion of the foundationalist picture of the structure of justification is hardly new. The logical positivist Otto Neurath—a member of the famous Vienna Circle gathering of intellectuals in the early twentieth century—famously suggested another metaphor. He likened justifying our beliefs to rebuilding a raft at sea. If we are to work on one of the planks, we must stand on another. If we later need to repair that second plank, then we must go back to standing on the first. We can't repair all the planks on a boat at sea at the same time. In other words, when we support our beliefs about one kind of thing, we take other beliefs for granted as justified. But we might later throw those into question, and take the first ones for granted. There is no point outside of the raft—outside our framework of beliefs—on which to stand.

A slightly more updated, but similar metaphor might be the "wiki." A wiki is a platform by which numerous people can participate in shaping a document or webpage. There is no single editor with a single "foundational" vision of how the work should turn out: changes are often made piecemeal, dropped in, replaced, reedited and so on. No single bit of information is immune to change. Or we might think of a fabric: Descartes understood knowledge to be secured by the strength of its foundations, but the fabric metaphor sees it being secured by the strength of its connections. What we know—if we are lucky enough to know— is woven together and constructed from many interlocking strands. Each of the strands supports the rest—some directly, some more remotely. In most weaves, no single thread or set of threads supports any of the others. The support they provide is, we might say, holistic, not linear. The same with spiderwebs. Or World Wide Webs.

The point of all these metaphors is the same. Webs, fabrics, interlocking planks of a raft and wikis are all networks, but they are not networks with foundational nodes; the nodes are where the individual lines and threads cross. And that, of course, is the point. Our beliefs are nodes in a network, supported by the over-all *coherence* of the fabric of beliefs to which they belong.

The "coherentist" picture of reasons does seem like a better description of how we justify our beliefs to one another in the Internet age. Nowadays, when we want to know whether something is true, we look it up on the Web. Practically speaking, that means checking to see how the relevant proposition hangs together with other things we think we know. Suppose I wanted to know the average size of sea turtles. I google it and find several pages that give me an answer. I pick Wikipedia. I then want to check whether Wikipedia is an accurate source of information on sea turtles. So I google that—and find that Wikipedia has an extensive page on whether Wikipedia is reliable (which is, in fact, the case—check it out). I may indeed find that my original belief about the average size of a sea turtle is justified; I find that it is confirmed by a page whose reliability is also confirmed. The whole pattern or structure of reasons here takes the form of an interlocking network.

As we saw in the first part of this book, however, knowledge comes in more than one form. That's crucial to remember right here, for the simple reason that only using Wikipedia to check on Wikipedia is circular. I've never really left the network of information I was consulting. In many cases, that is fine. If the circle of reasons is wide and big enough, we may not need to worry. But generally speaking, being trapped in a circle of reasons is—or should be—a disquieting fact. For it leaves open the possibility that our networks of reasons are just massive and mutually reinforcing fantasies. If our networks of reasons are really going to be justified, if they are really going to get us knowledge, then at some point they need to be anchored to something else, something beyond themselves. That is why it is a mistake—and I think here Weinberger and I might agree—to think that facts, and justifying our beliefs in light of them, are no longer important to the pursuit of knowledge. Giving up on the Cartesian dream of certain, immutable foundations doesn't mean that we should give up on anchoring our beliefs altogether.

How then *are* they anchored? In two ways. First, by the objective world itself—by what is true and what isn't. That's why, as I urged earlier, we don't want to give up on the idea of truth. The second is that reason-giving isn't all there is to knowing. We can also know by being receptive to the facts outside of ourselves, by having what the contemporary thinker Ernest Sosa calls "animal" knowledge or Descartes before him called *cognitio*.[15] That's a good thing to remember in this context. My network of reasons isn't just floating at sea. Some of the beliefs for which I have reasons are also ones that I know receptively, by responding to the environment in which I live with the senses I have. Others I may know receptively without being able to defend them with reasons.

Humans can get this kind of anchoring knowledge by getting up off the couch and plunging into the whirlpool of actual experience. It is still the best way, in my view—although not the only way—for us digital humans. To escape your circle of justification, do what you do with any circle: step outside its borders and breathe in the environment on the outside.

Of course for our receptive beliefs to be actually anchored to reality, reality must cooperate. And sadly, it often does not. That is why we are always forced back to look for reasons, to standards of reasonableness. We need assurance that the anchor is truly set. That gets us back to our network of reasons; we are back in the circle. As the philosopher Duncan Pritchard has noted, it is probably our lot as knowers to always be in some state of angst about our knowledge.[16] There is no getting around the fact that in order to know receptively, we have to be lucky; we either track the facts around us or we don't and are fooled again. The anchor sets or it doesn't.

I've argued in this chapter that the zeitgeist takes us to be networked knowledge machines. That's one of the lessons to draw from reflecting on the neuromedia thought experiment. As I've pointed out, knowledge and the process of justification is growing more networked in several ways: in its structure, in its source and, most radically, in the fact that our own cognitive capacities themselves are networked. In and of itself, this increasingly networked nature of knowledge isn't good or bad. It is just what is happening. What *can* be good or bad is how we react to this fact. As I've been urging, what we *don't* want to do is assume that because knowledge is networked, the nodes in the network—the individual knowers—no longer matter.

7.

Who Gets to Know: The Political Economy of Knowledge

Knowledge Democratized?

The Internet of Things and the networked knower are changing not only *how* we know, they are changing the *politics of knowledge*. And like all politics, the politics of knowledge is about power. In this case, it is the power over who gets to count as a knower and what gets to count as known. As Larry Sanger, philosopher and cofounder of Wikipedia, says, this is an awesome sort of power, because "it can shape legislative agendas, steer the passions of crowds, educate whole generations, direct reading habits and tar as radical or nutty whole groups of people that otherwise might seem perfectly normal."[1]

For much of Western history, it was the Church that determined what passed for knowledge. The means for exercising this power largely consisted in its ability to control who could read and what was written down—the Church both ran the universities and controlled the copying (by hand) of texts. Of course, after the print revolution, that began to change. The printing press allowed more people the opportunity to not only write down but mass-produce and distribute their own thoughts. Thus, what counted as knowledge became more diffuse, but also more accessible. Before long, however, power began to shift toward those who controlled the presses and means of distribution—and state imposition of copyright laws and censorship quickly became more prevalent and important. Since the eighteenth century, contemporary liberal societies have slowly (and not without much backsliding) made efforts to curtail state censorship and to allow ideas to spread more freely. Of course they too have had their own gatekeepers, even if their gates were more permeable: libraries, universities, publishers, the media. Yet as anyone who has been paying attention to these trends knows, those gates too have been coming down.

The Internet, it is often said, is democratizing knowledge. This is perhaps the single most heralded upside of the changes in informational technology we've been experiencing for the last two decades. But what does it mean to "democratize" knowledge—and how might current technologies contribute to that process?

First, and most obviously, the Internet, like the printing press before it, has made bodies of *knowledge more widely available*. The possibility of mass-produced books lowered the price at which knowledge could be bought and sold. As such, it brought such knowledge—and the possibility of literacy—to millions of people who had previously lacked access to it. Web 2.0 has greatly expanded this process while also changing both the sheer amount of different kinds of information available and the speed at which that information can be accessed.

A good example concerns this very topic. Try googling "How many people have access to the Internet," and sources such as Wikipedia and the International Telecommunications Union in Geneva will tell you that while roughly 94 percent of Swedes have Internet access, and 84 percent of Americans, only 2.1 percent of the population of Chad does. Nonetheless, the very availability of these statistics is a great example of the sort of information that just a few years ago you'd have had to go to a large research university to find or rely on journalists to report. While billions of people continue to have no access to it, millions have immediate access to the sort of information they wouldn't have had just a decade ago. In short: while it is far from ubiquitous, "more information to more people" is one obvious way that the Internet is making knowledge—or its acquisition—"more democratic."

The Internet is also democratizing knowledge by making its *production more inclusive*. One common example here is open source software like Mozilla's Firefox Web browser. When security vulnerabilities or bugs arise in Firefox software, a diverse and widespread community of volunteers works on fixes and plug-ins. Open source software operates similarly to an online co-op. It is software by the people, for the people.

Epistemic inclusivity is also a by-product of the growing number of open access research sharing sites such as Academia. edu. Founded in 2008, Academia.edu allows its millions of users (I'm one) a platform upon which to share and comment on one another's research. It allows researchers to pass their work *directly* to those who might be interested in it, or benefit from it.

Inclusivity of a different sort can come about by what *Wired*'s Jeff Howe dubbed "crowdsourcing" in 2006. Crowdsourcing is not simply any activity that uses the World Wide Web as a platform for people to network about problems—like Intrade or rankings on Amazon. As computer scientist Daren Brabham defines it, crowdsourcing is an online problem-solving and production model that "leverages the collective intelligence of online communities to serve specific organizational goals."[2] In other words, it is the top-down organized use of the Internet hivemind. An organization throws out a problem, and those who want (or those granted access to the relevant site or network) contribute solutions, and see what sticks.

The popular incentive-based innovation platform InnoCentive is often cited as an example of the inclusivity of crowdsourcing. (InnoCentive is ancient in Web 2.0 terms: it was founded in 2002.) Here's how it works: nonprofits and businesses post prize competitions for solutions to challenges. These can run the gamut—from retail product positioning to early detection mechanisms for inflammatory bowel disease. The prizes themselves vary in size, with some topping nearly a million dollars but many being significantly less. InnoCentive is only one example of how crowdsourcing can work, of course. Other famous examples include Amazon's Mechanical Turk, which allows companies (and scientific researchers) to outsource specific tasks to a huge network of "Turkers" to perform tasks that humans are still better at than computers, such as image identification and translation. Still another is Threadless, an organization that assigns a crowd of T-shirt designers the job of selecting (and creating) new T-shirt designs.

Challenge-specific prizes, like those used by InnoCentive, have been useful sources of innovation in scientific research for centuries. The British Crown spurred a huge leap forward in marine navigation in the seventeenth century, for example, by offering a prize for a device that could calculate a ship's longitude—resulting in the invention of the marine chronometer. Competitions like this work partly because they provide an incentive for "fresh eyes" on the problem. Indeed, researchers Lars Bo Jeppesen and Karim Lakhani, in their 2010 study of InnoCentive, suggested that there is an inverse relationship between a solver's likelihood of solving a problem and his or her degree of expertise in the field in question.[3] As Brabham writes, this means that, for example, "a biologist may fare better than a chemist would at solving a chemical engineering problem."[4] The same study also found that women significantly outperformed men as problem solvers on the site— despite, and possibly because of, the fact that they are often on the edges of the "scientific establishment." That is inclusivity of a very obvious sort.

A third way that the Internet has democratized knowledge is by making what is known more *transparent*—particularly with regard to information held by governments. The most obvious and controversial example of this is WikiLeaks, a nonprofit organization that publishes news leaks and classified governmental information online. Its disclosure of videos and documents related to the Iraq and Afghanistan wars in 2010 and 2011 caused a worldwide uproar. Supporters defended it as a tool for exposing the important facts that are relevant and needed in order for citizens to make informed democratic decisions. Critics denounced the organization as putting the lives of soldiers and diplomats at risk. Of course, both of these claims can be true—and whether or not WikiLeaks is ultimately beneficial or harmful, it is just the most visible example of the use of the Internet to enforce or encourage transparency. From revelations of NSA spying to videos of police mistreatment, the Internet can be used to shine a light on all sorts of activities, arguably empowering citizens.

So there is no doubt that the Internet has changed how we distribute, produce and reveal knowledge, and in many ways for the better. But using the language of "democratization" to describe these changes obscures as much as it describes. It ignores the fact that these changes aren't necessarily leading to more democratic ways of organizing our information society.

Epistemic Equality

Changes in how knowledge is distributed, produced and revealed indirectly affect the politics of knowledge—who gets to count as a knower and why—because they directly influence the *economy of knowledge.* By the economy of knowledge I mean, roughly, the structure of relations that divides epistemic labor and governs its exchange. The simple fact is that not all changes in the economy of knowledge—even those that can be legitimately described as "democratizing"—are leading *societies* to become more democratic.

Begin with a familiar economic bedtime story. Once upon a time, if you wanted a new chair, you either made it yourself or you went to a specialized craftsman. That craftsman would make the chair, but in turn received his materials and tools from still other craftsmen, and they in turn from others. In this way their labor—and expertise—was divided. No one person was responsible for the chair. A group—or a network of individuals, all of whom had some specialized knowledge—created it. Then (according to the story) one day the Industrial Revolution came, and the expansion of capital, and the invention of large machines to mass-produce chairs, allowed factories to create chairs without having to employ legions of expert furniture-makers. This lowered the price, despite the fact that the chair you bought from the factory was also produced by a network—indeed, an even larger network—with more specialized nodes. Some nodes were responsible for raw goods, as before, some for transportation, some for building special machine parts, and some for operating the individual machines that jointly produced the parts that another node would assemble into the chair. Then, one day (again according to the story), the global economy was born. And the network responsible for the chair you bought at Walmart got even bigger. It stretched across the globe, with the factory making the chair—or the iPhone on which you shopped for it—now in China, where the workers would be so specialized that some were hired simply for the size of their hands, and could be paid so little that the price of the products back home in the West could be cheaper than ever, despite the distance each product had to travel in order to find its way to your home.

Just as no one person can build everything, no one person can know everything. As a result, societies have always divided not only manual labor but intellectual labor between highly skilled laborers and expend significant public resources on training and rewarding those with such skills. That's precisely the way we still organize the medical, legal and scientific professions, for example. There is a network of knowledge, but it is a network in which the individual nodes are sources of expertise, and hence have a better chance of passing on good information and weeding out the bad.

The increasingly networked nature of knowledge challenges this model of the economy of knowledge. Indeed, the bestselling economist Jeremy Rifkin thinks that we are seeing the emergence of a new world order and the death of capitalism. We are instead, Rifkin suggests, seeing the rise of the Collaborative Commons:

> The IoT [the Internet of Things] enables billions of people to engage in peer-to-peer social networks and cocreate the many new economic opportunities and practices that constitute life on the emerging Collaborative Commons. That platform turns everyone into a prosumer and every activity into a collaboration . . . allowing social capital to flourish on an unprecedented scale, making a shared economy possible.[5]

As a result, Rifkin argues, the Internet of Things and the networked nature of our digital form of life are moving us toward "the zero marginal cost society." In turn, that challenges the central capitalist tenet, that increased human productivity requires increased human labor. "The traditional dream of rags to riches is being supplanted by a new dream of sustainable quality of life"—a life where we can spend more time engaged in pursuits that interest us, such as making music, cooking better food and thinking about philosophy.[6]

Rifkin thinks the same revolution is happening at the level of knowledge—perhaps especially there, since the wide availability of knowledge is the fuel powering the rest of our economy. But while Rifkin's collaborative vision might be appealing, the death of capitalism—and exploitative versions of it—is hardly near.

Take crowdsourcing as an example. Instead of an economy of skilled laborers that require resources to train, equip and compensate, crowdsourcing makes it possible for companies to distribute and generate knowledge without the expense of hiring those experts. This isn't necessarily more "democratic." But it is more capitalistic. Even some of the most active workers for Amazon's Mechanical Turk can make very little, two to five dollars an hour. This may seem reasonable if you think of such laborers as amateurs—doing such work in their "spare time." But as Brabham has convincingly argued, the idea of the "amateur crowd" is largely a myth. Turkers working for Amazon are generally highly educated professionals working in areas of the world where financially rewarding employment for those skills is significantly less than elsewhere (hence the attraction of Turk). In the case of InnoCentive, while it may be that nonspecialists are better solvers in many cases than those who self-identify as specialists in an area, this should not be taken to mean that the solvers are amateurs. Far from it: they are typically professional scientists. As Brabham sums it up: "these so-called amateurs are really outsourced professionals, and the products and media content that we are sold are not much different than the old products."

That's a key point. *Crowdsourcing is really a type of out-*

sourcing. And outsourcing knowledge production is as profitable as outsourcing anything else. It is simply a mistake to think that such outsourcing is making knowledge production more democratic. Indeed, the opposite seems to be the case: outsourced knowledge producers such as crowd workers are professionals without the protection of a profession—without, in short, basic labor rights. Crowd workers don't own what they produce. Indeed, as Brabham notes, in some cases, designers working "on spec" give up the rights to their designs, thereby forfeiting any future income from their intellectual labor. This is a win for the companies that employ such labor, but it hardly seems a win for democracy. With a large enough network, it doesn't matter if the individual nodes themselves have rights. If you have enough people, you still get similar results, and at a low wage. Cheap labor, good enough results. It is enough to make Sam Walton smile.

In short, the globalization of the economy of knowledge may be having some of the same effects as the globalization of the economy generally. One of the worst consequences of an unfettered, deregulated global economy is gross income inequality. This phenomenon isn't simply a matter of some people making more than others. It concerns a structural fact: that only a few control half of all global resources. It is part of a larger pattern of financial injustice But the unfettered global economy is not only increasing economic inequality, it is also encouraging *epistemic inequality.*[7]

To understand what I mean by epistemic inequality, let's think first about the value of equality itself. Equality, like liberty, is a core value of democracy, but it is often misunderstood. When

142

we say, with Locke and Jefferson, that all persons are "created equal," we aren't saying that we want everyone to be exactly the same, that we don't want diversity in abilities or talents. What we mean is that we are equal in our basic rights as individuals and, in particular, equal in having a claim on access to various resources. Thus the value of epistemic equality: the idea that *all persons have a basic claim to the same epistemic resources*. An epistemic resource is a structure or institution that provides information and at least the basis for knowledge. Thus, epistemic inequality is the result of an unfair distribution of structural epistemic resources.

The most obvious example of an epistemic resource is education. The United Nations holds education to be a fundamental human right. Arguably it is a basis for many other rights, or at least necessary for one to fully enjoy those rights. Without a basic education, people are unable to fully participate in contemporary societies (or almost any society): it is difficult to hold a job, access healthcare or make informed democratic decisions.

The rise of Web 2.0 has made the Internet a similar epistemic resource. Thus the UN has argued recently that preventing access to the Internet is itself a violation of fundamental rights. According to a UN special report, societies have an obligation to recognize the "unique and transformative nature of the Internet not only to enable individuals to exercise their right to freedom of opinion and expression, but also a range of other human rights, and to promote the progress of society as a whole."[8] Consequently, blocking that access is harmful. The concept of epistemic equality allows us to explain that harm directly. Removing access to the Internet, whether by criminalizing participation in online activities or explicitly blocking content, is wrong simply because it is an infringement of epistemic equality.

Epistemic inequality increases between groups when there is unequal access among those groups to epistemic resources: libraries, the Internet, education. The most obvious, and urgent, reason for this is poverty. Epistemic resources like libraries and Internet access come after food, shelter and health; without the latter, the former are unnecessary. This is a simple point but one often underemphasized. The set of digital human beings is not equivalent to the set of human beings, period. Most people on the planet are not participating in the Internet of Things, and many have never participated in the glories of Web 2.0. The Collaborative Commons is a first world dream. That doesn't make it a bad one. But predictions of the death of capitalism ignore the fact that much of the human population is exploited for its labor in order to make the 3-D printers and iPhones that we enjoy so much. And that fact isn't going away in a world where the black carbon soot emitted by small cooking fires is still a contributing cause of climate change.

Another cause of epistemic inequality is closed politics. A political society that is, roughly speaking, "open"—one that has a diverse and independent media, that protects freedom of information and communication and that exercises little government censorship—is apt to be more epistemically equal than one that is not as open. It is an ongoing question to what degree any society is truly open. But one thing is clear: the more closed a political system happens to be, the more apt the people in charge are to keep the epistemic resources to themselves. And as we saw in the last chapter, the more tempted they will be to abuse those resources at the expense of average citizens.

Even in societies that are relatively open in the political sense (that is, to the degree that political rights of expression and communication are protected), epistemic inequality can occur simply if the Internet is not relatively free and open. This is a third obvious cause for a rise in epistemic inequality. Because even if legally you can access whatever you like on the Internet in a given society, access is unequal if its cost is prohibitive or stratified by levels of service.

This is why the battle going on over Net neutrality is so important. *It is about epistemic equality*. Net neutrality is the idea that governments and Internet service providers should treat the information flowing through the Net equally. In particular, companies shouldn't be allowed to charge more for certain types of data. The argument on the other side is based on free market economics: when demand for a certain kind of traffic—say, Netflix or HBO GO—skyrockets, access to those services should cost more. At heart this is a debate about how to see the Internet. On one side are those who define the Internet as something that can itself be owned and profited from. On the other are those who feel it is an epistemic resource, like education or public libraries. In that case, if you start limiting access, you not only contribute to epistemic inequality, you contribute to inequality, period.

Issues around Internet access are also why, as Rifkin himself acknowledges, we should worry about the monopolization of the Internet. "What does it mean when the collective knowledge of much of human history is controlled by the Google search engine? Or when Facebook becomes the sole overseer of a virtual public square, connecting the social lives of 1 billion people?"[9] What it means is that the gatekeepers are back, and this time the gates, while now small, enclose more. And that, if we are not careful, could ultimately mean less epistemic equality.

So, even if knowledge is more "democratized" now—its production and distribution is more inclusive and available—that means little in conditions of increasing epistemic inequality. If you are too poor and oppressed to access anything online, the digital wonders of the world mean nothing to you. The value of epistemic equality is the value of open and fair access to epistemic resources. But "access" here means more than just the ability to go to school or look things up on the Internet. It also means something more abstract but just as important: *having the status of a full participant in the economy of knowledge.*

To be a full participant in a monetary economy, you need to be more than just a laborer. Slaves are laborers, but their labor is not shared or exchanged by them; it is stolen from them. To be a true economic participant, you need to be someone who has the resources and willingness to participate in buying and selling. But more than that: *you have to be recognized as such by others.* Otherwise, you end up just trading with yourself. Likewise with the economy of knowledge. To participate in that economy, you need to be more than just a receptive knower and reasonable believer. *You need to be seen or understood as such.* Otherwise your epistemic labor will be ignored or exploited. You won't be counted as a reasonable believer, as someone who can be trusted; you'll suffer what the philosopher Miranda Fricker labels "epistemic injustice."[10]

The history of racism in this country and many others is replete with examples of people being excluded from not only the monetary economy but the epistemic economy. In 1854, for example, the California Supreme Court infamously ruled that it was perfectly legal that "no Black or mulatto person, or Indian, shall be allowed to give evidence in favor of, or against a white man." In writing the opinion, Chief Justice Charles J. Murray pointed to what he thought was a slippery slope:

> The same rule which would admit them to testify, would admit them to all the equal rights of citizenship, and we might soon see them at the polls, in the jury box, upon the bench, and in our legislative halls. This is not a speculation . . . but an actual and present danger."[11]

Murray, for all his terrifying racism, sees the very point at issue. To recognize a class of people as possible testifiers in a court of law *is* a slippery slope—because it grants them the status of a reasonable believer. It treats them as credible participants in the economy, and as such, as persons who have autonomy over their thoughts and actions. That's one point that Fricker's work has brought to the fore in recent discussion: epistemic injustice of this sort has crippling effects. Once you are no longer recognized as a possible credible source of information—even about yourself—then the dominating class will excuse itself for ignoring your basic rights.

Epistemic injustice of this sort has been much discussed by writers in postmodern critical theory. But the general drift there has been to abandon the category of "reasonable" or "justified" belief—to see these as inherently dominating categories. What is interesting and important in Fricker's work is that she doesn't see it this way. For her, abandoning standards of reasonableness would be giving up on the goal of epistemic equality. In short, what we need is not to abandon reasonableness but instead, in philosopher Lewis Gordon's words, to "shift the geography of reason."[12] And that is the question we would be wise to ask with regard to our digital life as well: how is it contributing to, or inhibiting, that shift? A central cause for worry is the increasing fragmentation of reasons themselves. In the context of our present discussion, we might worry that this fragmentation doesn't just have bad political effects. It has bad epistemic effects. It promotes epistemic inequality and a loss of intellectual autonomy. And that in turn can affect people's ability to filter out bullshit—simply because their filtering is so one-sided.

Web 2.0 and the Internet of Things *can* be forces for democratic values. But we must not let our enthusiasm blind us to the existence of epistemic inequality, and the fact that its causes—racism, income inequality—pollute the infosphere just as much as they pollute the minds that make it up.

Walmarting the University

Standard procedure for university exams these days involves prohibiting the use of smartphones. As I was reminding my students of this recently, one of them joked that the university better come up with policies on wearable tech like smart watches ASAP. We all laughed, but nervously, because he was right. And as another student noted, whatever policy that is, it is going to be outmoded by the time it is enacted—not just because universities are slow to adapt to change but because technology is moving so fast. While Google's initial experiment with Glass may not have been successful, the idea isn't going away; and one day that too may seem quaint, should something like neuromedia emerge.

That raises a question: if the Internet is available to you at the blink of an eye—and available in a way that seems like memory—then what *are* we testing for when giving exams? What, in general, is the point of higher education in the age of big data?

These questions come at a time when the idea of the university itself is often said to be in crisis—especially in the United States. In one sense, the American university system continues to flourish. American institutions of higher learning dominate world rankings, making up more than half of the top 100 and a large majority of the top ten. Go to any top research conference in the world and you'll find many of the keynote speakers and top researchers there are from American universities. American institutions continue to lead in the production of scientific research in the best journals, and produce the most Nobel laureates. And students from across the world continue to come to the United States to study. In economic terms, university education continues to be one of America's leading industries.

But at the same time, there is the increasing worry that we are in something of an education bubble, and that the model is no longer sustainable. The cost per university student for an education has risen almost five times the rate of inflation since 1983. Thus it is not surprising that the amount of debt per student has so dramatically increased; two-thirds or more of students now take out loans.[13] Private institutions routinely charge around $60,000 a year, and an "affordable" public institution, like my own, can cost more than $25,000. The explanations for these depressing facts vary, although it is clear that part of the matter is that state funding, on whom both public and, to a lesser extent, private institutions have long depended, has dramatically decreased in the last three decades.[14] Taxpayers, for good or for ill, no longer clearly favor paying for the epistemic equality brought about by public institutions—and public education, at all levels, is obviously a primary victim of this change in mentality. But whatever the explanation, it is hard to avoid the conclusion that something needs to change.

Starting around 2012, many pundits, and more than a few academic administrators, started forecasting that information technology was going to lead this change. In particular, the advent of MOOCs (Massive Open Online Courses) was thought to signal a shift to a different model of education. MOOCs are free (or mostly free) online courses, composed generally of video lectures, various forms of computer-enabled discussion forums and computerized grading. In the wake of several high-profile successes attracting thousands of students, startups and non-profits promoting and hosting MOOCs, such as Coursera and edX, sprang up almost overnight. Universities began creating their own MOOCs. The anticipation, and the hype, ran high, with the president of edX, Anant Agarwal, declaring that MOOCs would reinvent education and transform and "democratize education on a global scale."[15]

MOOCs do indeed have much to offer. Many of the courses

allow people who would never have a chance to take a course by a world-renowned expert on a subject the ability to do so, and for free too. In many cases, students can even receive college credit if they finish the course successfully. Already millions of people around the globe have taken advantage of this opportunity. As a result, it is hard not to see it as crashing the gates of the university and helping to promote epistemic equality. It is also simply edifying, as a friend of mine (a superstar teacher who designed and created a MOOC while at the University of Virginia) said to me. Few things are more inspiring than to find yourself talking philosophy to 80,000 people worldwide, from all levels of income and backgrounds. Who can argue against free philosophy?

Not me. Yet only two years later, it is becoming clearer that, for all their many virtues, MOOCs are not exactly the revolutionary product they have been hyped to be. To see why, let's go back to Rifkin. According to Rifkin, the old model of higher education maintained that "the teacher was akin to the factory foreman, handing out standardized assignments that required set answers in a given time frame."[16] The old model was "authoritarian" and "top-down." It emphasized lectures, was hierarchical in its power structure and privileged memorization over discussion. The new model emerging in the Collaborative Commons is more lateral, egalitarian and interdisciplinary.

Rifkin is certainly right that the old, old, *old* model of education has many of the features he describes. But the *Mad Men* era has been gone for some time now, and the shift to more discussion-orientated, problem-solving models of education began as far back as Dewey. And this was the result not of a technological shift but a pedagogical one. This helps explain why many educators have been skeptical about using MOOCs as a replacement for, as opposed to an addition to, brick-and-mortar classroom teaching. *Most MOOCs, after all, just are paradigm examples of the old model in action.* They consist of lectures. Their methods of assessment are standardized. They privilege memorization over discussion. While those are not essential features of MOOCs, of course, the technology is only as innovative as we want it to be, and, right now, it seems as if we don't want it to be that innovative. The fact that MOOCs are more like big lectures is why faculties at Amherst and Duke have rebelled against involving their institutions in MOOCs. Their point was not that there is something inherently wrong with making education free online—far from it. Their point was that the present models of MOOCs are simply extensions of what is already happening at universities worldwide: large classroom lecture-style courses. Pedagogically, many (although not all) MOOCs are not innovative; they are old school.

The other reason educators have been wary of MOOCs is that some see them as hastening what we might call the Walmarting of the university. As I noted above, a hallmark of the global economy is cheaper goods, produced and sold by poorly compensated workers, made possible by amazing models of distribution. This trend has been dominating education as well. According to a leading study of the American professoriate, in 1969 over three-quarters of faculty at American colleges and universities were in reasonably well-paid and stable tenure-track positions.[17] By 2009, that number had almost flipped, with only about one-third of faculty now being tenure-track. In short, most students are now taught by temporary workers who are largely not union-ized and paid well below the minimum wage. The worry that many have had about MOOCs is that it will only exacerbate this process, should universities (as some initially proposed to do) replace their own course offerings with MOOCs purchased for their students from other entities.

Whether that will come to pass is hard to say. MOOCs are in their infancy, and their path is hard to predict. But it's doubtful that MOOCs are the biggest changes looming in education due to technology. Instead, those changes will likely come more directly via the Internet of Things. As I stated at the outset of this section, the big questions concern the more obvious fact: what do you make of education when people have all the "facts" at hand? If you had neuromedia, you'd be able to access tons of information about history, philosophy, mathematics, art, etc. You'd have dates and names at your disposal, just as you do now on your phone. You'd Google-know all sorts of stuff—that is, you'd have potential receptive knowledge, as I've put it. And the more Google-knowledge we have, the greater the "room" in stu-dents' minds, you might think, for more important stuff.

This isn't only a modern problem. The use of technology to outsource mental activities is hardly new. At one time, calcula-tors were verboten in math classrooms; not any more. Similarly, students today routinely access the Internet during instruction, and often do so in an interactive way designed and monitored by the instructor. (I've done this in my own courses.) Let's also remember that libraries have long provided huge riches of knowl-edge for those who want them. Thus the question, "Why go to college if you have neuromedia?" is not much different than the question (one I took seriously myself as a know-it-all youth), "Why go to college when you have a library?"

You already know the answer to that one. In the ideal world, if not always the reality, we go to college to find pilots who can guide us across the vast seas of knowledge. We need them to tell us what is already charted and what is left to chart still. Such guides shouldn't make us more receptive knowers; they should aim to make us more reflective, reasonable ones and, what's more, they should help us to understand.

8.

Understanding and the Digital Human

Big Knowledge

Google knows us so well that it finishes our sentences. This program, known to any user of the Internet, is called Google Complete. Search as I just did for "Web 3.0 and . . ." and Google will suggest "big data" and "education"; search for "knowledge and . . ." and you might get "power" and "information systems." Complete is a familiar, if rather gentle, form of big data analysis. It works because Google knows not only what much of the world is searching for on the Web, but also what *you've* been searching for. That data is useless without Google's propriety analytic tools for transforming the numbers and words into a predictive search. These predictions aren't perfect. But they are amazingly good, and getting better all the time.

Google has done more than perhaps any other single high-profile company or entity to usher in the brave new world of big data. As I noted in the first chapter, the term "big data" can refer to three different things. The first is the ever-expanding volume of data being collected by our digital devices. The second is analytical tools for extracting information from that data. And the third is the firms like Google that employ them.

One of the lessons of previous chapters is that big data and our digital form of life, while sometimes making it easier to be a responsible and reasonable believer, often makes it harder as well—while at the same time setting up conditions that make reasonable belief more important than ever before. The same thing could be said for understanding—except even more so. And that's important, because understanding is what keeps the "human" in what I earlier called the digital human.

The End of Theory?

In 2008, Chris Anderson, then editor of *Wired,* wrote a controversial and widely cited editorial called "The End of Theory: The Data Deluge Makes the Scientific Method Obsolete." Anderson claimed that what we are now calling big data analytics was overthrowing traditional ways of doing science:

> This is a world where massive amounts of data and applied mathematics replace every other tool that might be brought to bear. Out with every theory of human behavior, from linguistics to sociology. Forget taxonomy, ontology, and psychology. Who knows why people do what they do? The point is they do it, and we can track and measure it with unprecedented fidel-

ity. With enough data, *the numbers speak for themselves.* . . . Petabytes allow us to say: "Correlation is enough." We can stop looking for models. We can analyze the data without hypotheses about what it might show. We can throw the numbers into the biggest computing clusters the world has ever seen and let statistical algorithms find patterns where science cannot.[1]

Traditional scientific theorizing aims at model construction. Collecting data is just a first step; to do good science, you must *explain* the data by constructing a model of how and why the phenomenon in question occurred as it did. Anderson's point was that the traditional view assumes that the data is always limited. That, he says, is the assumption big data is overthrowing.

In 2013, the data analytics expert Christian Rudder (and cofounder of the dating website OkCupid) echoed Anderson's point. In talking about the massive amount of information that OkCupid (and other) dating sites collect, Rudder writes:

Eventually we were analyzing enough information that larger trends became apparent, big patterns in the small ones, and even better, I realized I could use the data to examine taboos like race by direct inspection. That is, instead of asking people survey questions or contriving small-scale experiments, which was how social science was often done in the past, I could go and look *at what actually happens*, when, say, 100,000 white men and 100,000 black women interact in private.[2]

Anderson and Rudder's comments are not isolated; they bring to the surface sentiments that have been echoed across discussions of analytics over the last few years. While Rudder has been particularly adept at showing how huge data gathered by social sites can provide eye-opening correlations, and data scientists and companies the world over have been harvesting a wealth of surprising information using analytics, Google remains the most visible leader in this field. The most frequently cited, and still one of the most interesting, examples is Google Flu Trends. In a now-famous journal article in *Nature*, Google scientists compared the 50 million most common search terms used in America with the CDC's data about the spread of seasonal flu between 2003 and 2008.[3] What they learned was that forty-five search terms could be used to predict where the flu was spreading—and do so in real time, as they did with some accuracy in 2009 during the H1N1 outbreak.

Google Flu Trends—which we will look at again below—is really only an extension of the design methods behind Google's main search engine. Its algorithms (and their creators) don't know why one page is what you want rather than another; they just apply mathematical techniques to find patterns in incoming links. That's all. Similarly, Google Flu Trends doesn't care why people are searching as they do; it just correlates the data. And Walmart doesn't care why people buy more Pop-Tarts before a hurricane, nor do insurance companies care why certain credit scores correlate with certain medication adherences; they care only that they do. As Viktor Mayer-Schönberger and Kenneth Cukier put it, "predictions based on correlations lie at the heart of big data. Correlation analyses are now used so frequently that we sometimes fail to appreciate the inroads they have made. And the uses will only increase."[4]

Does the use of big data in this way however, really signal the end of theory, as Anderson alleged? The answer is no. And, as we'll see, that is a very good thing.

Start with Rudder and Anderson's remarks. As Rudder puts it, big data seems to allow us to investigate by direct inspection. We don't have to look through the lens of a model or theory; we can let the numbers speak for themselves. Big data brings us to the real-life correlations that exist, and because those correlations are so perfectly . . . well, *correlated*, we can predict what happens without having to worry about why it happens.

But can we *ever* look at the "data in itself" without presupposing a theory? In *The Structure of Scientific Revolutions*, Thomas Kuhn famously argued that you cannot: data is always "theory-laden." His point was that there is no direct observation of the world that isn't at least somewhat affected by prior observations, experiences and the beliefs we've formed as a result. These beliefs in turn set up expectations. In short, theory permeates data.

This operates even at the level of deciding what experimental techniques or devices to employ. As Kuhn put it, "consciously or not, the decision to employ a particular piece of apparatus and to use it in a particular way carries an assumption that only a certain sort of circumstances will arise. There are instrumental as well as theoretical expectations."[5] In support of the claim, Kuhn cited the now-classic 1949 article by Bruner and Postman on perceptual incongruity. Bruner and Postman showed their subjects playing cards, some of which had abnormalities (a spade card was red, for example).[6] What they found was that, primed with ordinary cards, respondents identified the abnormal cards as perfectly normal; their expectations seemingly affected what they saw. The last seventy-five years of psychology have only underlined the lesson (if not necessarily the letter) of Bruner and Postman's experiment. What you believe can affect what you observe.

Rudder and Anderson may well protest that they don't mean to deny Kuhn's point. They aren't worried about perceptual observations but mathematics. But even when it comes to mathematical correlations detected by mindless programs, our theoretical assumptions will matter: they will determine how we interpret those correlations as meaningful and, most importantly, what we do with them.

A trivial example of how assumptions can matter in this way occurs in Rudder's book. When discussing a well-known data map that tracks the "emotional epicenter" of an earthquake by looking at Twitter reactions, Rudder notes that "Knowing nothing else about a quake, if it were your job to distribute aid to victims, the contours of the Twitter reaction would be a far better guide than the traditional shockwaves around an epicenter model."[7] Maybe so; and the data map, and others like it, certainly are interesting. But Rudder's point here rests on some key assumptions. First, it assumes that aid workers won't be concerned about aftershocks (which will be better predicted by models employing geological and geographical data). Second, it assumes that all types of quakes generate equally explicable Twitter reactions. (What happens, for example, if people are too injured to type?) Third, it not only assumes that people have equal access to smartphones, but that their first priority is to tweet rather than rescue the injured. In an extreme quake, the emotional epicenter as charted by Twitter may be far away from the point of truest need. My point is not to overhype what is a passing remark in a much longer work; it is to show that data correlations themselves are useful only under certain background assumptions. And where do those assumptions ultimately come from? *Theory.*

Another word for this is *context.* Without it, correlations can be as misleading as they are informative. A recent and extremely striking example is art historian Maximilian Schich's video map of cultural history (reported in *Science,* with a following video posed by *Nature*).[8] Schich and his colleagues, employing data gleaned by Freebase (a huge set of data owned by, who else, Google), used the mathematical techniques of network analysis to map what they referred to as the development of cultural history. After collecting data about the locations and times of the births and deaths of 150,000 "notable" people over 2,000 years, they made a video map of the data (with births in blue and deaths in red). What resulted showed how, over time, "culture" moved and migrated—sometimes, it seems according to the map, clustering around certain cities (Paris, a center of red) and sometimes more widely distributed. The video is arresting (if you haven't seen it, google "Schich and cultural history"). The idea, Schich said, was to show that you could do in history what is done in the sciences: use data to show actual correlations rather than relying on armchair theorizing.

But Schich's data map relies on a host of assumptions. A good deal of discussion of the video on Twitter and elsewhere following its release concerned the Eurocentric nature of the map. The notable figures chosen by Schich were almost entirely white, European and male (and in many cases, possessing some wealth). This makes the widely viewed video—which talks about cultural history *simpliciter*—not just striking but strikingly cringeworthy at points. In fairness, Schich was well aware of this bias; the researcher's point, as he noted, was to use the available data to discover patterns in broadly European cultural history.

Yet Schich's assumptions don't stop at race, gender and ethnicity, nor are they all the products of available data. Some of his assumptions are about how to define "culture." Schich's map suggests that culture is driven by notable figures (from scientists to movie stars). But is what used to be called "the big man" theory the only or best way to understand what shapes cultural change? What about economics or politics, for example? Other assumptions concern how the drift of culture is measured. Why think that where someone died has more predictive value for cultural development than where they spent their most productive years? Descartes, for example, died in Sweden, but he spent most of his productive life in France. Once again, theoretical assumptions drive work in big data as much as they do in any other field. Kuhn would not be surprised.

None of this diminishes the importance of network analyses as tools for research, including fields typically not associated with data, like history. It's a growing and exciting mode of research across anthropology, literature, the digital arts and the humanities. But as the historian and digital humanities scholar Tom Scheinfeldt has remarked, this work is only as good as the theoretical context in which we place it.[9]

So, like it or not, we can't do data analytics without theory. It's what gives us the context in which to pose questions and interpret the correlations we discover. But we should like theory; the process of theorizing employs a composite of cognitive capacities, ones that when employed together make up *understanding*, another way of knowing that is important to human beings.

Understanding Understanding

Suppose you want to learn why your apple tree is not producing good apples. You google it and the first website you look at (for example, the ACME Apple Research Center) is a source of scientific expertise on apples. It tells you the correct answer, call it X. But there are many other websites (e.g., the nefarious Center for Apple Research) that came up during your search that would have told you the wrong answer, and many others (e.g., MysticApples.com) that would have given you the right answer but for the wrong reasons. So you could have easily been wrong about X or right about it but for the wrong reasons.[10]

Silly as it is, this example replicates how we now know much of what we know, as I've been pointing out in this book. We know by Google-knowing. Not that there is anything wrong with that. After all, in the above case, we are being responsible and believing X, based on an investigation and on the basis of a reliable source.[11] In several ways, then, you could be said to know X. And for most purposes, that's good enough. Yet it is clear that something valuable can sometimes go missing even when you go about the process responsibly. Sometimes we need to know more than the facts; sometimes we want to understand. And it is our capacity to understand that our digital form of life often undersells, and which more data alone can't give us.

Understanding is a complex form of knowing, one that has several facets or elements. The first is that understanding isn't piecemeal; it involves seeing the whole. For example, think about the difference between knowing a pile of individual facts about some subject, theory or person and actually understanding that subject or theory or person. Understanding involves knowing not just the facts, but also the *how* or *why* something is the case. You understand more about the Civil War if you understand why and how it came about; you understand string theory if you understand why it predicts certain events; you understand a person to the extent you don't just know that she is unhappy, but what makes her unhappy. In each of these cases you are going beyond mere data to grasp something deeper and more profound.

The philosopher Stephen Grimm, who has thought as much about this topic as anyone recently, has pointed out that there is something in common between understanding *how* something is and *why* it is.[12] In both cases, he argues, we "grasp" or "see" not just individual elements, but the *structure* of the whole. This sounds grand, and it can be, as when we understand how a particular equation works or why a great historical event occurred. But it can also happen on a smaller scale. Consider, for example, the lucky person who understands how her car works. She has this understanding in part because she has certain skills, skills that give her the ability to see how various parts of a mechanism depend on one another: you can't get the car to move without the battery and the battery won't be charged without the alternator.

You understand when you see not just the isolated bits, but how those bits hang together. Similarly with understanding why. When we understand why something is the case, such as why a certain disease spreads or why your friend is unhappy, or why a given apple tree produces good apples, we grasp various relationships. These relationships are what allow us to see the difference between possibilities, between one hypothesis and another.

So, understanding is a kind of knowing that involves grasping relationships—the network, or parts and whole. But crucially, the relationships you grasp when you understand something aren't just correlations. To truly understand, you also need to know what *depends* on what—*why* the spread of a certain disease is related to hand-washing habits, or why having good apples depends on having a certain amount of rainfall.

The dependency relations we grasp when we understand can come in different forms. Some relations might be about cause and effect. Think of a game of chess: if I move my bishop to a certain square, I cause it to change its position. But they might also be logical: if I move my bishop to this square, it will be vulnerable to your pawn. Or semantic: the bishop can move to that square because the rules define it as being able to move diagonally across the board. In other words, the first important element of understanding is grasping dependency relations: having systematic knowledge of how things both fit together and depend on one another, causally, logically and otherwise.[13]

That's why the person who truly understands something is often the person who can best explain it. In Plato's famous dialogue *Euthyphro*, the title character is an unlikable busybody and self-anointed expert in religious matters. Socrates meets him on the steps of the courthouse, where Euthyphro is on his way to prosecute his own father, of all people. As it happens, Socrates is waiting for pre-trial proceedings to begin in his own trial for impiety and blasphemy against the gods. Socrates "innocently" asks Euthyphro to tell him what holiness is—he could use some advice, he says. Euthyphro answers, rather fussily, that holiness is what is loved by the gods.

No doubt, says Socrates slyly. But which comes first? Do the gods love what is holy because it is so, or is something holy simply because the gods happen to love it? Socrates is pointing out that Euthyphro's answer is really just expressing a correlation:

x is holy when, and only when, x is loved by the gods.

If Euthyphro is right, being holy is perfectly correlated with being loved by the gods. So if you know what the gods love, you can perfectly predict what is holy—and vice versa. But this leaves it open which side of that correlation is really doing the work. It leaves open what depends on what. As such, it really doesn't offer an explanation of *why* the gods love what they do. Hence, it also doesn't offer an explanation of *what* holiness is.

Plato's point is instructive for another reason: it shows that this need for explanation can arise even in cases where the question of how much data (or how little) is moot. Euthyphro's equation, after all, purports to be a perfect correlation, and as such would be ideal for prediction. But it still doesn't answer the question. And were you an ancient Greek, it would be a question that would matter for how you conceived of your relationship to the gods. If you think of the gods as discovering what is of ultimate value, then you think ultimate value is more fundamental to the universe. If you think they create it, then you think the gods are more fundamental. Which way you go, as Socrates later hints, can change how you see your relationship not just to the gods but to the universe itself.

Understanding is the kind of knowledge you need in order to be able to give a good explanation of something. And this is why we think of explanations as involving more than mere correlation. They make us aware of why things hang together, which in turn allows us to see that understanding is a matter of degree— the larger and more coherent the set of information one has about apple trees, etc., and the greater one's reflective and intuitive awareness of the coherent connections between one's beliefs about those matters, the greater one's understanding.[14] The greater your grasp of the whole, the better able you are to fully explain the phenomenon in question.

Knowing How to Chuck

We've talked about Plato. Now let's go highbrow and talk about Chuck. *Chuck* is a TV comedy in which the title character downloads a huge amount of secret NSA information from an "Intersect" computer. In the second season, a 2.0 version of the Intersect machine comes out, and Chuck is suddenly able to do more than just download facts. The computer literally drops abilities—like the ability to be super-good at kung fu—right into him. He then becomes (as the show says) the government's "most valuable asset." Hilarity ensues.

Chuck, in short, is imagined as having gained knowledge from something like neuromedia. But the knowledge he supposedly gains isn't just of the book-learnin' type. It is know-how. He somehow acquires a skill, or what psychologists call procedural knowledge.

Understanding has a complex relationship to procedural knowledge, or the knowledge involved in having skills. This relationship is important not only for understanding understanding, but for grasping the extent to which technology like neuromedia can help us truly understand and how it cannot. So let's think about skills and procedural knowledge for a minute.

Over the course of the twentieth century, the dominant view of procedural knowledge has been that there is a very sharp difference between knowing how and knowing facts. The Oxford don Gilbert Ryle, in his influential 1949 book *The Concept of Mind* sets himself against the idea that "the primary exercise of minds consists in finding the answers to questions."[15] Knowing how to do something, Ryle suggests, isn't a matter of knowing a particular fact. Instead, it is more like having an ability to do something. And the philosopher Hubert Dreyfus has influentially argued that knowing how to do something—like ride a bike—can't simply be understood as grasping a set of rules or directions. At bottom it involves a type of discernment or acuity that can't be discursively codified.[16]

The idea that there is a sharp difference between knowing how and knowing facts seems to have some empirical support as well. Consider the famous case of the patient HM. HM was an epileptic who had undergone a lobectomy. He was then observed to have severe anterograde amnesia. In other words, he would forget events almost immediately after they happened. In a groundbreaking experiment, the cognitive psychologist Brenda Milner had HM perform a mirror-drawing task in which he had to draw the outline of a star through a mirror while not being able to see his own arm.[17] The results were astounding: he was able to improve at the task after several days, even though he had no memory of the event. And that may suggest that acquiring a skill is completely distinct from having knowledge of facts, since he got better at doing something that he could never remember having done before.

This interpretation—and the sharp difference between practical and theoretical knowledge that goes with it—has been recently challenged. As the philosopher Jason Stanley and neuroscientist John Krakauer point out in a recent paper, HM *was* given explicit instructions before performing the mirror task. He was able to use that knowledge. Of course, being HM, he later forgot that he had that knowledge. He wasn't able to articulate it. But that doesn't mean he didn't possess the knowledge at any point, and that it didn't play a causal role in his ability to engage in the drawing. More generally, Stanley and Krakauer argue that motor skill tasks involve not just the manifestation of a motor acuity—the ability to make discriminations—but also the employment of receptive knowledge of facts. Motor skills, in other words, are package deals—they are complex states "requiring both increasing knowledge of required actions, and practice-related improvement in the selection and acuity of these actions."[18] In other words, knowing how to ride a bike, or how to play tennis, involves not just physical acuity but at least some, probably unconscious, discursive knowledge.[19]

One lesson we can learn from both perspectives—despite their differences—is that greater mastery of a skill involves more knowledge of the complex sort I've been calling *understanding*.

Aristotle and Plato saw the relationship between what the Greeks called *epistêmê* and *technê*. For the Greeks, *technê* meant an organized body of know-how or procedural knowledge—cooking, farming, sailing, knitting, programming and playing jazz are all examples. But for Plato and Aristotle both, really mastering a *technê* or skill meant that the expert understood—in the sense I've been explaining—the craft.

For the Greeks, the true expert—whether that person is a craftsman or a scientist—is someone who understands what they are doing. That understanding is what allows them to say why good apples differ from bad, to explain how a streamlined computer program works, to articulate the difference between the good political policy and the disastrous.[20] That's why mastering a skill, for the Greeks, was not the same as having good habits, or even just having a knack or talent.[21] To truly master a skill, you of course need some talent, serious motivation, lots and lots of experience and practice; but you also need to understand how the details fit together, how the parts add up to something greater than themselves.[22]

The reciprocal relationship between skills and understanding is partly why experts can seem so baffling to the novice. When I was a young man, I studied martial arts with an instructor who drilled me repeatedly in certain traditional forms of striking and throwing. These forms involved following what were strict rules of movement. But when I saw my teacher spar with more advanced students, he would often deviate from those forms. When I asked him why he was not following the very forms I was being told to learn, he replied that he *was* following them. I just didn't know enough yet to understand how various actions fit together, and how the skilled practitioner, to fit the moment, could modify a form. The master of a skill both knows the form, knows how it fits into the big picture, and has a certain acuity—one that, as Dreyfus says, cannot be reduced to a list of discrete discursive knowledge. He has an understanding that allows him, as Aristotle might say, to see how the universal is present in the particular.

So, one way in which understanding is related to skills is that mastering a skill produces understanding. But understanding of any type—understanding, as such—also essentially involves the *manifestation of a particular set of skills.*

Let's consider another story about Socrates, who was supposedly told by the Oracle of Delphi that he was the wisest man in Athens. Socrates famously replied that he only knew that he knew very little, or what he didn't know. So what sort of knowledge did he have? Well, consider what he was truly good at. One thing, surely, was asking questions. This came from a combination of facts and the ability to draw connections between them. As a result he had *know-which*, as it were. He knew *which* questions to ask.

Knowing-which is another element of understanding. The person who truly understands, in the philosophical sense, is discerning not only the actual situation, but also why various hypotheses and explanations *won't* work as well as how to ask what would. They know why kicking the refrigerator *here* and not *there* will help get it working. This is something that Socrates was great at, and it is something that experts in general can do. It also explains why understanding increases the better one is at asking the right questions. Experts—those who understand a given subject best—are often able to increase their understanding even further because they have the ability to know which question they should ask in the face of new information. By so doing, they can not only reveal that Euthyphro knows nothing of piety, but that the good folks at MysticApples.com know hardly anything about apples.

Being able to ask a good question, however, is not the core point. That's because the skill of being able to ask good questions itself hinges, at least in part, on a simpler (and less overtly verbally orientated) cognitive capacity: the ability to make inferences and draw out a position's consequences—and not just the actual consequences of, say, a given position on what causes apples to be tasty, but also the consequences of that position in certain counterfactual situations. This is precisely the skill that a good doctor employs when considering whether to administer a drug, or a lawyer uses when considering an argument. It is also, arguably, the skill a good mechanic employs when considering whether to disassemble a head gasket, or an apple farmer uses when deciding whether another farmer's advice is reasonable. And those who have the capacity to cognitively engage, should they have the requisite verbal and linguistic abilities, will know which questions they should ask in order to carry their inquiries even further.

Understanding, then, what the Greeks called *epistêmê*, is a multifaceted sort of knowledge. It involves knowing why and how but also knowing which—which questions to ask, and where we might go next. As such, it both comes from and involves procedural knowledge—skills.

Procedural knowledge generally comes from experience. You get skills through practice, through a relationship with the world. That's not to say you can't gain theoretical knowledge—knowledge of facts—from direct experience. It is probably the best and fastest way in most cases. It is just that you can also gain knowledge of facts by reading. But acquiring a *skill* requires at least some experience. You can't *just* read about it. You need to do something, to practice, to try things out, to fail, to start again.

People can be trained to do something they couldn't do before. That is what we are probably thinking about when we suspend belief about poor Chuck. Drugs, after all, provide a temporary type of programming for one's body. And drugs can *change* your abilities—make you stronger, faster, less prone to depressive thoughts. But taking a drug to become faster is fundamentally different from learning how to skillfully run faster. Mastering a skill requires trial and error. Such experimentation is how one develops the perceptual and informational acuity—the ability to discriminate between what works and what doesn't—that is part of the package deal of knowing how to do something well. And that is why Chuck isn't (I know you'll be shocked to hear this) the most coherent of fictions. We might be able to acquire a certain level of skill by downloading—just by acquiring new basic abilities. But downloading knowledge other people have acquired via experience isn't the same as having that experience yourself, isn't the same as personal trial and error and creative adaptation in the face of circumstance. Downloading, in short, won't give you mastery. And it won't give you the understanding that comes with it.

To truly understand some things, you need to develop a skill; and skills require experience. Which means that understanding often does as well. That's a point that many of us already grasp intuitively. One sees this in the wellspring of interest in the last few years in organic foods, home brewing and back-to-the-farm movements, especially amongst those under thirty-five—the very demographic of people most heavily invested in, and used to, information technology. The underlying explanations for such movements are of course complex, but part of the story is a shared recognition that doing something yourself—making or growing something—gives you a type of understanding that you would lack otherwise. That's also why many parents want their children to participate in activities that require hands-on experience. As one parent expressed it to the *New York Times*, "My partner and I saw that kids were spending too much time interacting with perfect interfaces. . . . We felt that we needed to provide an experience by which they could understand how perfection is achieved—and, more specifically, how that perfection is achieved by working through problems with your hands."[23]

The relationship between experience and understanding tells us something important about the limits not only of neuromedia but of Google-knowing. Google can give us the world. And the Internet and Web 2.0 can certainly give us the information that we need to learn new skills—gaming skills, skills at Web interfacing. These skills can be useful in certain situations, and not just online. But we kid ourselves if we think that we can learn every skill we need simply by downloading it. We need to interact with the world outside our head to do that.

Coming to Understand as a Creative Act

Descartes was a late riser. His habit, when possible, was to stay in bed till around noon—musing. One day, according to legend, he was watching a fly zoom around above his head when, suddenly, he realized that he could track its position by measuring its distance from the walls and the ceiling. He understood how to plot its flight path in space . . . and voilà! We get Cartesian coordinates, or so the story goes.

The story of Descartes' fly—and others like it, such as those about Newton's apple or Einstein's clock—are instructive because they emphasize that the moment of understanding can involve sudden insight. Such moments are often called "aha moments" and, in the psychological literature, are collectively taken to signify the "Eureka effect" (so named after Archimedes, who after a moment of great insight shouted "Eureka!"). Of course, most acts of understanding do not require the sudden novel inspiration that Descartes had. But all of them do involve some level of insight. Having such an insight is part of why understanding is fundamentally a creative act.

Coming to understand is a mental act in the same way that reflecting or deciding are mental acts. They are activities that your mind engages in. They take effort and increase the total cognitive load. Don't confuse the state of understanding and the act of coming to understand. One can be in a state of understanding, just as one might be in a state of decision (or indecision, as the case may be), without doing anything in particular, or even being conscious of being in that state. Your understanding in such a case is tacit or implicit. Much of what we understand we understand in this way (consider: you probably understand why the water you put in your freezer turns to ice, but you didn't think about it until just now). But in order to understand, one must first *come to understand*, and it is this coming to understand that is an act. It is no coincidence that even the terms we use for it, such as "grasping," are often active.

To talk about creativity, or creative acts, is to open a Pandora's box of multifarious treasures that can soon get away from us. So let me just say what I mean by it here: a mental act is creative to the extent that it generates novel and valuable ideas. As Margaret Boden, the cognitive scientist and AI researcher, has emphasized, creative ideas needn't be historically novel—like Descartes' new geometrical ideas—but they are psychologically novel to the creator.[24] Thus, being creative isn't the same as being original. People can have ideas that are creative *for them*. As Boden says, "Suppose a twelve-year-old girl, who'd never read *Macbeth*, compared the healing power of sleep with someone knitting up a raveled sleeve. Would you refuse to say she was creative, just because the Bard said it first?"[25] I don't think so, and neither does Boden. Creativity is relative to a person.

But creativity is not just novelty. If that were so, then too many thoughts would satisfy the criterion of being "creative" to make it worth talking about. Creative ideas are valuable to the person's cognitive workspace. They move things forward on the conceptual field on which they are currently playing. They are useful and fecund. They have progeny, and they contribute to the problems at hand.

Creative acts are also surprising in a certain sense. In cases of sudden insight, this leads to the "eureka" feeling. But creative acts can be surprising even if they do not necessarily provoke that "aha" feeling. Boden calls this their "impossible" aspect— that is, an idea is creative for a person when we *affectively experience it as novel*, when from the inside, it feels like it could not have been had prior to the moment of creation. Conditions were right, and the person suddenly "sees."

Coming to understand why or how something is the case is a particular kind of creative mental act in the sense that I just described. That's because, paradigmatically, it involves generating new, valuable and surprising ideas. Which ideas? Those that concern dependency relationships—how things fit together. The "grasping" of those relationships, which lies at the heart of understanding, is what makes understanding creative.

This may seem most obvious in the paradigmatic, historic cases of understanding, like Descartes' geometrical insight or Einstein's flash of understanding relativity upon seeing a clock. But what about less historically original acts of understanding? Consider again a child who comes to understand, for the first time, why 0.150 is smaller than 0.5. At that moment, the child is also having an insight—a realization of how things are related. Or consider again our student above, coming to understand for the first time why Lady Macbeth sees blood on her hands, or why sailing is more pleasant and efficient when the wind is not behind you. Each of these acts of understanding are creative insights for the person in question, even though they are in no way original.

They are also surprising—again, not necessarily in the "eureka" sense—because the person who comes to understand could not, relative to their past evidence and cognitive context, have understood it before that moment. If understanding is creative, then it is both active and passive. That's because the surprising or "impossible" aspect of creativity makes creating seem at once something we do (which it is) and at the same time something happening to us. The muse suddenly strikes. Realization comes in a flash. Understanding is like this as well. It involves insight, and insight, as the very word suggests, is the opening of a door, a "disclosing," as Heidegger said. One acts by opening the door, and then one is acted upon by seeing what lies beyond. Understanding is a form of disclosure.

9.

The Internet of Us

Technology and Understanding

I began this book by pointing out a paradox: our digital form of life seems to both expand and inhibit our knowledge, simultaneously. How can that be?

As I've argued throughout, the first step is to see that "knowing" does not name a single kind of cognitive process, except in the minimal sense that to know is to have a grounded sense of what is true. To know can mean being receptive, or being reasonable, or understanding. Yet that is only part of the story. The second point I've stressed is that our digital form of life tends to put more stock in some kinds of knowing than others. Google-knowing has become so fast, easy and productive that it tends to swamp the value of other ways of knowing like understanding. And that leads to our subtly devaluing these other ways of knowing without our even noticing that we are doing so—which in turn can mean we lose motivation to know in these ways, to think that the data just speaks for itself. And that's a problem—in the same way that our love affair with the automobile can be a problem. It leads us to overvalue one way to get where we want to go, and as a result we lose sight of the fact that we can reach our destinations in other ways—ways that have significant value all their own.

When it comes to knowing in the receptive sense, our knowledge is radically extended beyond ourselves. By virtue of the technology in our pockets, on our wrists and in our glasses, you and I are already sharing information-producing processes. We are cognitively interconnected by the strings of 1s and 0s that make up the code of the infosphere. That is the truest sense in which knowledge is more networked now, and why it is not an exaggeration to say, as Jeremy Rifkin does, that the Internet "dissolves boundaries, making authorship a collaborative open-ended process over time."[1] In turn, this raises the possibility that digital humans' receptive abilities are not only more networked but that our acts of understanding may be becoming more networked as well.

In one really obvious sense, information technology is helping us understand more than ever before. That's because we also know in the receptive sense more than ever before. Google-knowing is a terrific *basis* for understanding in the way that reading a textbook is. You can't connect the dots if you don't have the dots in the first place. Moreover, neuromedia, and even existing digital media, increases our ability to make connections between bits of information. That's helpful to understanding, since understanding increases with inferential and explanatory connections between beliefs.

Yet Google-knowing, while a basis for understanding, is not itself the same as understanding because it is not a creative act.

To use the Internet is to have the testimony machine at your fingertips. That is what makes it so useful. But understanding is often said to be different from other forms of knowledge precisely because it is not *directly* conveyed by testimony—and thus not directly teachable.[2] Again, you can give someone the *basis* for understanding. But in the usual cases, you can't directly convey the understanding itself. An art teacher, for example, can give me the basis for creative thought by teaching me the rudiments of painting. She can give me ideas of what to paint and how to paint it. But I did not create these ideas; I create when I move beyond imitating to interpret these ideas in my own way. Likewise, you can give me a theorem without my understanding why it is true. And if I do come to understand why it is true, I do so because I've expended some effort—I've drawn the right logical connections. Coming to understand is something you must do for yourself.

Let's contrast this with other kinds of knowledge. I can download receptive knowledge directly from you. You tell me that whales are mammals; I believe it, and if you are a reliable source and the proposition in question is true, I know in the receptive way. No effort needed. Or consider responsible belief: you give me some evidence for whales being mammals. You tell me that leading scientists believe it. If the evidence is good, then if I believe it, I'm doing so responsibly. But in neither case do I thereby directly understand why whales are or aren't mammals. You can, of course, give me the explanation (assuming you have it). But to understand it, I must grasp it myself.

Or so it is generally. One might wonder, however, whether that would remain the case were we as fully integrated as the neuromedia possibility imagines. To have neuromedia would be like reading minds. You'd be able to access other people's thoughts through little more than the intermediary of satellites. We would all be Google Completing our thoughts for one another, and as result collaboration could very well start to *feel* from the inside like individual creation does now.

This is still a long way from showing that neuromedia would increase our understanding of the world all by itself. There is no doubt that information technology is already radically facilitating collaboration. And coming to understand, like any act of creation, is something you can do *with* others. But just because you can understand with others doesn't alter the fact that understanding involves a *personal* cognitive integration—a combination of various cognitive abilities in the individual, including a grasp of dependency relations and the skill to make leaps and inferences in thought. It ultimately involves an element of individual cognitive achievement. *Understanding is not something I can outsource.*

Yet what makes this individual cognitive achievement so valuable? Why worry about understanding if correlation, as Chris Anderson might say, gets you to Larissa? What can it add that other forms of knowing cannot?

Understanding is a necessary condition for being able to explain, and explanations matter. A well-confirmed correlation can be the basis of (probabilistic) predictions. But prediction is not the only point of inquiry, nor should it be. Good explanations for why a correlation holds give us something more. As the eminent philosopher of science Philip Kitcher has noted, good explanations are fecund.[3] They don't just tell us what is; they lead us to what might be: they suggest further tests, further views, and they rule out certain hypotheses as well. Moreover, if you want to control something and not just predict what it will do given the preexisting data, you need to know why it does what it does. You need to understand. Thus, being able, on the basis of Google Flu Trends, to predict where the flu spreads is incredibly helpful. But if we want to know how to control its spread, we must better understand *why* it spreads. And once we do so, it seems likely that our predictions might themselves become more nuanced.

In fact, authors of a recent study critiquing the predictive power of Google Flu Trends have made this very point.[4] The authors argue that more refined predictive techniques drawing on traditional methods of modeling can be at least as accurate as Google's method, which they demonstrate has routinely overestimated the amount of flu cases by as much as 30 percent. They ascribe this to what they call "big data hubris," or the assumption that sheer data size alone will always result in more predictive power. The researchers' point is not that big data techniques aren't helpful, but that the Google algorithm is not likely to be a good stand-alone method for predicting the spread of the flu.

Given our argument above, this is not surprising. Big data techniques are going to assist our models and explanations, not supplant them.

The creativity of understanding helps to explain our intuitive sense that understanding is a cognitive act of supreme value and importance, not just for where it gets us but in itself. Creativity matters to human beings. That's partly because the creative problem-solver is more apt to survive, or at least to get what she wants. But we also value it as an end. It is something we care about for its own sake; being creative is an expression of some of the deepest parts of our humanity.

Finally, understanding can also have a reflexive element. Our deepest moments of understanding reveal to us how we ourselves fit into the whole. Thus, an act of understanding something or someone else can also help you understand yourself. When that happens, understanding comes with what Freud called the "oceanic feeling"—the feeling of interconnectedness.

Perhaps this is why we treasure those moments of understanding in both ourselves and others. If you've ever taught or coached or parented someone, you've tried to help someone understand. The moment they do is what makes the effort worthwhile. If that moment never comes, you regret it because that person is missing out on an act of creative personal expression, a chance to see how the parts connect to make the whole.

So even if, contrary to what I've suggested here, we are someday able to outsource our understanding to some coming piece of glorious technology, it is not clear that we should want to. To do so risks losing something deep, something that makes us not just digitally human, but human, period.

Information and the Ties That Bind

What would it be like if you had the Internet connected directly to your brain? That, or something like it, is the future toward which we are barreling. The hyperconnectivity of our phones, cars, watches and glasses is just the beginning. The Internet of Things has become the Internet of Everything, the Internet of Us.

These pages have spun a cautionary tale about this progress, but there is actually a lot to be optimistic about. The massive amount of data that is making hyperconnected knowing possible has the potential to help cure diseases, contribute to constructive solutions to climate change and tell us more about our own preferences, prejudices and inclinations than we ever thought possible. I look forward to these developments, and I hope you do too. My point in this book is that we should nonetheless approach the future with our eyes wide open, especially since our relationship with the Internet is becoming more and more intimate. Intimacy brings comfort, but it also makes us vulnerable.

Some of these vulnerabilities are extensions of those we already have. The Internet of Us will be comprised of human bodies that are themselves communicating with one another, and with the Net, through a variety of embedded or surface-worn devices. Data trails will follow us around like so many little sparks; dancing points not of light but of 1s and 0s. These data trails are already here. I am reminded of Aleksandr Solzhenitsyn's remark in his 1968 book *Cancer Ward*:

> As every man goes through life he fills in a number of forms for the record, each containing a number of questions. . . . There are thus hundreds of little threads radiating from every man, millions of threads in all. If these threads were suddenly to become visible, the whole sky would look like a spider's web, and if they materialized as rubber bands, buses, trams and even people would all lose the ability to move, and the wind would be unable to carry torn-up newspapers or autumn leaves along the streets of the city. They are not visible, they are not material, but every man is constantly aware of their existence. . . . Each man, permanently aware of his own invisible threads, naturally develops a respect for the people who manipulate the threads.[5]

The threads are strings of information. They are the ties that bind us to one another, and society to us. What big data and the hyperconnectivity of knowledge are doing is making these connections brighter, more numerous, stronger and fundamentally easier to pluck. And so our respect—if that is the word—should also grow for those who have, or wish to have, their hands on these strings. Let us hope their motivations are pure, or at least neutral, while we stay on guard for the opposite. As Bertrand Russell once remarked in a somewhat different context, advances in technology never seem to bring along with them—at least, all by themselves—a change in humanity's penchant for greed and power. That is a lesson I hope we heed—even while we look forward to the benefits the Internet of Us will bring.

Many of us share the same concerns. After the initial launch of Google Glass, the reaction was more negative than expected. While many were excited about the technology, it seemed that just as many were worried about its potential for invading privacy; others were concerned about its potential for distracting drivers. These practical objections were serious. But I can't help wondering if the concern went deeper. Before its launch, Google cofounder Sergey Brin was reported to have said, "We started Project Glass believing that, by bringing technology closer, we can get it more out of the way."[6] Brin was meaning to emphasize the fact that Glass allows you to take pictures without fumbling for your camera. But he inadvertently put his finger on a more basic fear of the Internet of Us. We are getting technology out of the way by pulling it closer—in the case of Glass, literally making us *see* through it. We know technology can always alter our perspective. But this perspective-altering effect can only increase as it migrates inward.

We must be careful that we don't mistake the "us" in the Internet of Us for "everything else." The digital world is a construction and, as I've argued, constructions are real enough. But we don't want to let that blind us to the world that is not constructed, to the world that predates our digital selves. And the Internet of Us is not only going to affect how we see our world; it will affect our form of life. One aspect of this concerns autonomy. The hyperconnectivity of knowledge can help us become more cognitively autonomous and increase what I called epistemic equality. But I've argued it can also hinder our cognitive autonomy by making our ways of accessing information more vulnerable to the manipulations and desires of others. And it can lead us to overemphasize the importance of receptive knowing—knowing as downloading.

Humans are toolmakers, and information technologies are the grandest tools we have at the moment. Our tool-making nature shapes how we understand the world and our role within it. It encourages us to see the natural environment as something upon which we operate, which we use as means for our own ends, as an extension of the tools we develop to interact with it. So what happens when we extend our tools to the point that they become integrated with our life, when we become the very tools themselves? That is the most salient question about the coming Internet of Us. And it raises a danger that we cease to see our own personhood as an end in itself. Instead, we begin to see ourselves as devices to be used, as tools to be exploited.

None of this is inevitable, however. How could it be—the changes in our form of life that digital ways of knowing are bringing have yet to fully unfold. We should not fear information technology per se, or the "Internet" in the expanding Internet of Us. It is the "us" part—or our *uses* of technology—that we must mind. We *are* becoming more powerful knowers. We just must also strive to be more responsible, understanding ones.

Acknowledgments

Over the years, I've been fortunate to talk about these subjects with many wise and intelligent people, including Robert Barnard, Don Baxter, Paul Bloomfield, Sandy Goldberg, Patrick Greenough, Hanna Gunn, Julian Jackson, Casey Rebecca Johnson, Brendan Kane, Junyeol Kim, Nathan Kellen, Tom Lynch, Helen Nissenbaum, Nikolaj Jang Lee Pedersen, Duncan Pritchard, Baron Reed, David Ripley, Paul Roberts, Marcus Rossberg, Evan Selinger, Nate Sheff, Tom Scheinfeldt and Daniel Silvermint. A special shout-out to the Block Island Cognitive Research Institute, who heard early versions of these ideas (over and over again): Paul Allopenna, Terry Berthelot, James Dixon, Inge-Marie Eigsti, Lisa Holle, Jim Magnuson and Emily Myers.

Nate Sheff and David Pruitt were of great help in researching various materials in the initial stages of this project. Early drafts of the manuscript benefited heavily from comments by Patricia Lynch, Phil Marino, Kent Stephens, Tom Stone and Steven Todd;

Acknowledgments

Terry Berthelot provided invaluable commentary on a later draft. Portions of this book were given as talks at the University of Connecticut Humanities Institute, the University of Edinburgh, the University of St. Andrews, Northwestern University's Kaplan Humanities Institute, University of Cincinnati's Taft Center, Syracuse University, Ohio State University, the American Philosophical Association, Yonsei University, TEDx, the Chautauqua Institution and SXSW. Portions of chapters 4 and 6 build on ideas I first tried to express in "A Vote for Reason," "Privacy and the Concept of the Self" and "Privacy and the Pool of Information" in the *New York Times'* The Stone blog, as well as "The Philosophy of Privacy: Why Surveillance Reduces Us to Objects," May 7, 2015, in *The Guardian*. The ideas of chapter 1 draw inspiration from "NeuroMedia, Knowledge and Understanding," published in *Philosophical Issues: A Supplement to NOÛS*, vol. 24 (2014).

Finally, I owe special thanks to my agent Peter Matson and my editor Phil Marino, who both believed; editor Allegra Huston, who clarified; my sisters Patty, Bridget and Rene, who taught; and to Terry and Kathleen, who not only understand, but help me do the same.

Notes

Preface

1. Russell, "The Expanding Mental Universe."

Chapter 1: Our Digital Form of Life

1. So far, most research focusing on BCIs concerns how to get information from the brain to the computer in order to bring about
some external effect—such as controlling a robotic arm, or moving a cursor or an object like a wheelchair. Ongoing research is
geared toward making prosthetics or providing therapy by directly
stimulating the brain, although there are also people working on
using BCIs for gaming. A chief problem facing such research is
that signals from the brain—typically recorded noninvasively
with electrodes attached to the scalp—are quite noisy. It is hard
to filter out the information you really want. (One researcher

compared it to listening to a conversation in a football stadium from a blimp.) Nonetheless, advances are ongoing; in September 2014, an international team of researchers announced the first successful brain-to-brain verbal communication using BCIs. The researchers reported being able to send thoughts of words (in this case "Ciao" and "Hola") to other people's brains over the Internet without typing on a keyboard—just by thinking. See Grau et al., "Conscious Brain-to-Brain Communication."

2. For more examples of cyborging, and some related thought experiments, see Rose, *Enchanted Objects,* 23ff. See also Clark, *Natural-Born Cyborgs.*

3. Levy, *In the Plex,* 67.

4. For discussion, see Brian Merchant, "With a Trillion Sensors, the Internet of Things Would Be the 'Biggest Business in the History of Electronics,'" *Motherboard,* October 29, 2013. http://motherboard .vice.com/blog/the-internet-of-things-could-be-the-biggest-business -in-the-history-of-electronics. Accessed September 4, 2015.

5. Rifkin, *The Zero Marginal Cost Society,* 11.

6. Mayer-Schöneberger and Cuker, *Big Data,* 9.

7. Floridi, *The Fourth Revolution.* 25–58.

8. Cavell, *Must We Mean What We Say?,* 52.

9. Wittgenstein, *Philosphical Investigations,* 226.

10. Recent books reflecting these themes include Weinberger, *Too Big to Know*; Bilton, *I Live in the Future*; Rifkin, *The Zero Marginal Cost Society*; Rudder, *Dataclysm.*

11. James, *Pragmatism.* 164–65.

12. Wieseltier, "Among the Disrupted."

13. A few other sticks have been placed in the stream, for example: Carr, *The Shallows*; Roberts, *The Impulse Society*; Sunstein, *Republic.com 2.0.*

14. Shannon, "A Mathematical Theory of Communication." See Gleick, *The Information*, for discussion.

15. Grimm, "Is Understanding a Species of Knowledge?"; Kvanvig, *The Value of Knowledge*, 185–96.

16. Bostrom, "Are We Living in a Computer Simulation?" Bostrom's argument is ingeniously simple: Assume that in the future, some culture of super beings eventually reaches technological "maturity." That means they can, among other things, make SIM programs that, unlike current SIM programs, are completely lifelike—that would even have, as it were, *conscious* SIMs. Second, assume that these future super beings would want to run such programs—maybe they are curious how people in the past might have lived, or maybe just to amuse themselves. Assume too that if they ran one such program, they could, and would, run millions. If they do, the math of the situation is clear. Each SIM program would have billions of SIM lives. There are millions of SIM programs. If so, the set that would include all the "real" people that ever lived would be much smaller than the set of all SIMs. Thus, given the assumptions, it is more likely that you are in the bigger set than in the smaller. It is more likely that you are a SIM.

    This is an interesting result, in part because the argument is so simple. Whether we should be worried, of course, depends on how much stock we are willing to put in the assumptions: that technological maturity will happen before we all kill one another, or the sun burns out, or whatever; and that future people will be interested in running millions of SIM programs with billions of SIMs each. Moreover, as Bostrom is perfectly up-front about, there is the question of whether SIMs could ever be conscious in the way we are conscious, not to mention able to coherently consider the thought that they are.

Chapter 2: Google-Knowing

1. As the philosopher Hilary Kornblith has noted, it is consistent with this thought that we may not need to appeal to receptivity to explain the cognitive behavior of *particular* organisms. If all I wanted to explain was why some particular bird—a plover, say— leaves its nest and thrashes about an area while moving away from the nest, I need only appeal to its "belief" that there is a predator nearby. Whether there really is a predator nearby is irrelevant. But when it comes to explaining the capacities of *species*, things are more complicated, precisely because we are interested in how these capacities are *adaptive*. See Kornblith, *Knowledge and its Place in Nature*, 53–55.

2. I say in most cases, but it is possible, even plausible, that in some cases our cognitive capacities are actually spandrels. That is, cer- tain capacities—the capacity for abstract thought might be an example—are by-products of selective pressures aimed in a very different direction.

3. Reliability in this sense is much discussed in contemporary episte- mology, where seminal texts on the topic include Goldman, "What is Justified Belief" and *Epistemology and Cognition*. In the sense intended, a mechanism will be more reliable the higher the ratio of accurate to inaccurate information it produces over a given time. Just how high the ratio is will vary, and it will be environmentally dependent. For some organisms, mechanisms and environments, a given organism will be perfectly successful along some measure by employing an informational mechanism that is reliable at a ratio only slightly greater than chance. In other cases, a significantly higher degree of reliability will be imposed by environmental demands. Thus, for example, a given bird's visual mechanisms must

be very reliable at detecting moving objects if it is to succeed in capturing its prey. Yet the same bird's ability to discern stationary discrete object-pairs might only be slightly better than chance.

4. Kahneman, *Thinking, Fast and Slow*, 105.

5. Ibid., 86.

6. Chabris and Simons, *The Invisible Gorilla*, 5.

7. Sunstein, *Republic.com 2.0*, 69–90.

8. Locke, *An Essay Concerning Human Understanding*, 58.

9. Kant, "An Answer to the Question: What Is Enlightenment?"

10. Descartes, *Meditations*, 12.

11. My description of this problem is influenced by Edward Craig's description of the "epistemic state of nature" in *Knowledge and the State of Nature*, 11–13.

12. Put another way: reasonable belief matters because epistemic trust matters. Yet this is consistent with reasonable belief having value in its own right. After all, one might think that believing reasonably is simply intrinsically good, in the same way one might think it is intrinsically good to act in morally responsible ways. This would be natural if you thought of reasonable belief as grounded in the intellectual virtues.

13. See Lynch, *In Praise of Reason*, 79–88, for a more detailed presentation of this point.

Chapter 3: Fragmented Reasons

1. Popper, *The Open Society and Its Enemies*, vol. 1, 187.

2. For a useful version of this definition, and discussion of the themes of this chapter in general, see Coady, *What to Believe Now*, 120.

3. For some influential examples, see Putnam, "Bowling Alone," and Turkle, *Alone Together*.

4. Some researchers have recently argued that online communities can indeed build social capital—if under certain conditions. See, for example, Sajuria et al., "Tweeting Alone."

5. For a discussion of this, see Pew Research Center, "Political Polarization and Media Habits," and Iyengar and Hahn, "Red Media, Blue Media."

6. Sunstein, *Republic.com 2.0*, 69–83.

7. Ibid.

8. Yardi and Boyd, "Dynamic Debates."

9. Hemingway, "Lies, Damned Lies, and 'Fact-Checking.'"

10. A. Peter Galling, "Do Creationists Reject Science?," February 4, 2008. Available at: http://www.answersingenesis.org/articles/2008/02/04/do-creationists-reject-science. Accessed August 25, 2015.

11. Philosophers typically take ancient skeptical arguments as challenges to the possibility of knowledge. Whether they are depends on what kind of knowledge you have in mind. They don't challenge what I've called receptive knowledge, since all that is needed for such knowledge is belief formation that is reliable in fact. What they challenge is the possibility of reflective knowledge and giving reasons for what we believe to those who see things differently. They challenge reasonableness. See Wright, "Scepticism and Dreaming"; Pritchard, *Epistemic Luck*; and Sosa, "How to Resolve the Pyrrhonian Problematic," for similar views.

12. Hume, *Enquiries Concerning Human Understanding*, 272–73.

13. Hazlett, "The Social Value of Non-Deferential Belief," 9.

14. Rawls, *Political Liberalism*, xvii.

15. See also Uhlmann et al., "The Motivated Use of Moral Principles"; Graham et al., "Mapping the Moral Domain," 366.

16. Haidt, *The Righteous Mind*, 89. Haidt's fascinating and perceptive book concerns much more than the points focused on here; its principal aim is to diagnose the causes of ongoing political rifts.

17. Haidt, "The Emotional Dog and Its Rational Tail," 820–25.

18. Ibid., 817. I don't meant to suggest, and neither does Haidt, that such feelings *can't* be defended.

19. See Kahneman, *Thinking, Fast and Slow*, 41, and the view of Flanagan and Williams in "What Does the Modularity of Morals Have to Do with Ethics?" On the following point about changes, see Paxton et al., "Reflection and Reasoning in Moral Judgment."

20. Haidt, "Reasons Matter."

21. For an overview, see: http://www.apa.org/about/policy/parenting .aspx.

22. Adam Liptak, "In Battle over Gay Marriage, Timing May Be Key," *New York Times*, October 26, 2009.

23. Bloom, "The War on Reason"; Bloom, "How Do Morals Change?"

24. Mercier and Sperber, "Why Do Humans Reason?," 59.

25. As Nathan Kellen reminds me, even here it is difficult, as debates over the continuum hypothesis (the idea that there is no set of real numbers whose size or cardinality is intermediate between the reals and the naturals) show.

26. Haidt, *The Righteous Mind*, 85.

27. Ibid., 310.

Chapter 4: Truth, Lies and Social Media

1. "The 'Truth' Deleted from Internet in China," Daily Telegraph, July 13, 2002.

2. Stanley, How Propaganda Works, 54.

3. https://answersingenesis.org/dinosaurs/when-did-dinosaurs-live/ what-really-happened-to-the-dinosaurs/. Accessed August 25, 2015. This alarming result has been noticed by a number of science bloggers as of this writing.

4. Floridi, *The Fourth Revolution*, 50.

5. Ibid., 43.

6. For excellent explanations of social construction, see Haslanger, "Ontology and Social Construction" and *Resisting Reality*. Here I am concerned with the construction of both intentional and non-intentional objects.

7. Flanagan, *Dreaming Souls*, 134.

8. The most famous being Robert Nozick, in *Anarchy, State, and Utopia*, 40-41.

9. It is worth emphasizing that this line of reasoning is not intended to show, absurdly, that we want all of our actual beliefs to be true. I believe many propositions that I don't want to be true. Beliefs about the future of global warming or the continuing spread of AIDS in Africa number among them. But the fact that I don't want these particular propositions to be true is entirely consistent with it being the case that I care about believing what is true and only what is true, whatever that turns out to be. See Lynch, *True to Life*, 17–18.

10. Anecdotal evidence suggests that the younger you are, the longer you'd be willing to try a super-SIM life—and one would predict that trend to reverse at a certain age. That is, one would expect older, healthy, and reasonably happy people to be willing to try out the SIM life, but for a shorter period of time (again, other things being equal: persons of any age who are miserable or unhealthy might well sign up for SIM lives permanently).

11. D'Agata and Fingal, *The Lifespan of a Fact*, 107.

12. Nick Fielding and Ian Cobain, "'Revealed: Us Spy Operation That Manipulates Social Media'", *Guardian*, March 17, 2011.

13. Ian Urbina, "I Flirt and Tweet. Follow Me at #Socialbot," *New York Times*, August 10, 2013.

14. Silverman, *Verification Handbook*.

15. Weinberger, *Too Big to Know*, 112.

16. Achinstein, *The Nature of Explanation*, 69.

17. Ron Suskind, "Faith, Certainty and the Presidency of George W. Bush," *New York Times Magazine*, October 17, 2004.

Chapter 5: Who Wants to Know

1. Priest and Arkin, *Top Secret America*, 75.

2. Rifkin, *The Zero Marginal Cost Society*, 75–77.

3. Scalia (dissenting), *Maryland v. Alonzo King, Jr.*

4. For an in-depth discussion of some of the complexities here, I recommend Nissenbaum, *Privacy in Context*, 67ff. See also Lane et al., *Privacy, Big Data and the Public Good.*

5. Barton Gellman, Julie Tate, and Ashkan Soltani, "In NSA-intercepted Data, Those Not Targeted Far Outweigh Those Who Are," *Washington Post*, July 5, 2014.

6. http://www.bloomberg.com/news/2013-08-23/nsa-analysts-intentionally-abused-spying-powers-multiple-times.html. Accessed August 24, 2015.

7. See this ruling: http://www.nytimes.com/interactive/2013/08/22/us/22nsa-opinion-document.html, 16, n. 14. Accessed August 25, 2015.

8. http://www.whitehouse.gov/sites/default/files/docs/2013-12-12_rg_final_report.pdf.

9. As of 2015, the Office of the Director of National Intelligence's official report was still allowing significant, if more limited, incidental collection. See: http://icontherecord.tumblr.com/ppd-28/2015/overview. Accessed August 25, 2015.

10. Bloustein, "Privacy as an Aspect of Human Dignity," 974.

11. Ibid., 973.

12. Strawson, *Freedom and Resentment*, 9.

13. Sue Halpern, "The Creepy New Wave of the Internet," *New York Review of Books*, November 20, 2014.

14. One example is the introduction of contrary counsel. For some of the complications facing such a proposal, see this report from the Congressional Research Service: "Reform of the Foreign Intelligence Surveillance Courts: Introducing a Public Advocate," available at: http://fas.org/sgp/crs/intel/R43260.pdf. Accessed August 25, 2015.

Chapter 6: Who Does Know

1. Nietzsche, "On Truth and Lie in an Extra-Moral Sense," 47.

2. Weinberger, *Too Big to Know*, 45.

3. Caldarelli and Catanzaro, *Networks*, 16.

4. Clark and Chalmers, "The Extended Mind." See also Clark, *Natural-Born Cyborgs*, and Clark, *Supersizing the Mind*.

5. Goldberg, *Relying on Others*, 79ff.

6. Gilbert, *Joint Commitment*, 23ff. See also Gilbert, *Sociality and Responsibility*.

7. You might think that one way of understanding Gilbert's view—namely, that groups can literally have *beliefs* that are completely independent of their members' beliefs—is that it is a thought too far; compare it to U.S. presidential nominee Mitt Romney's infamous claim that "corporations are people too." It is not literally true. Perhaps all the examples show is that individuals, in so far as they are group members, can be implicitly or tacitly committed to a proposition that no group member is committed to as an individual. After all, the interviewer example presumes that the tacit commitment in question is itself a product of the individuals' beliefs. If so, then perhaps the best we can say is that what we can loosely call the group's implicit commitment "supervenes" or is a product of the individuals' commitments.

8. Surowiecki, *The Wisdom of Crowds*, xii.

9. David Leonhardt, "When the Crowd Isn't Wise," *New York Times*, July 7, 2012.

10. Nate Silver, "The Virtues and Vices of Election Prediction Markets," *New York Times*, October 24, 2012.

11. I was helped to see these points in discussions with Sandy Goldberg and Nate Sheff. The example in the text is similar to that in Goldberg, "The Division of Epistemic Labor," 117.

12. Weinberger, *Too Big to Know*, 21.

13. Descartes, *Meditations*, 103.

14. Weinberger, *Too Big to Know*, 23.

15. Sosa, *Reflective Knowledge*, chs. 7 and 8..

16. Pritchard, *Epistemic Luck*, 225.

Chapter 7: Who Gets to Know

1. Lawrence M. Sanger, "Who Says We Know: On the New Politics of Knowledge," *Edge* 208 (April 25, 2007): http://edge.org/3rd_culture/sanger07/sanger07_index.html%3E. Accessed August 25, 2015.

2. Brabham, *Crowdsourcing*, xix.

3. Jeppesen and Lakhani, "Marginality and Problem-Solving Effectiveness."

4. Brabham, *Crowdsourcing*, 21.

5. Rifkin, *The Zero Marginal Cost Society*, 18.

6. Ibid., 19. See also 179–80.

7. Fricker rightly distinguishes epistemic inequality from what she calls epistemic injustice: *Epistemic Injustice*, 1–2. But the two are related, as noted below.

8. Frank LaRue, Special Rapporteur on the promotion and protection of the right to freedom of opinion and expression, Report to

the Human Rights Council of the United Nations General Assembly, May 16, 2011. Available at: http://www2.ohchr.org/english/bodies/hrcouncil/docs/17session/A.HRC.17.27_en.pdf. Accessed August 28, 2015.

9. Rifkin, *The Zero Marginal Cost Society*, 204.

10. To be precise, what she calls "testimonial" epistemic injustice. See Fricker, *Epistemic Injustice*, ch. 2.

11. *People v. Hall.*

12. Gordon, "Shifting the Geography of Reason" and *Disciplinary Decadence*.

13. "Higher Education: Not What It Used to Be," *Economist*, December 1, 2012.

14. Michael Mitchell, Vincent Palacios, and Michael Leachman, "States are Still Funding Higher Education at Pre-Recession Levels." Center on Budget and Policy Priorities, May 1, 2014. Available at: http://www.cbpp.org/research/states-are-still-funding-higher-education-below-pre-recession-levels?fa=view&id=4135. Accessed August 28, 2015.

14. Carole Cadwalladr, "Do Online Courses Spell the End for the Traditional University?," *Guardian*, November 10, 2012.

15. Schuster and Finkelstein, *The American Faculty*, 40. See also Introduction.

16. Rifkin, *The Zero Marginal Cost Society*, 109.

Chapter 8: Understanding and the Digital Human

1. Chris Anderson, 'The End of Theory: The Data Deluge Makes the Scientifc Method Obsolete," *Wired* 16, no. 7: June 23, 2008.

2. Rudder, *Dataclysm*, 10–11.

3. Ginsberg et al., "Detecting Influenza Epidemics Using Search Engine Query Data."

4. Mayer-Schöneberger and Cukier, *Big Data*, 55–56. The examples just above also come from this interesting and informative book.

5. Kuhn, *The Structure of Scientific Revolutions*, 59.

6. Bruner and Postman, "On the Perception of Incongruity."

7. Rudder, *Dataclysm*, 196.

8. Schich et al., "A Network Framework of Cultural History."

9. In conversation.

10. This example illustrates an everyday experience for all of us. But it also illustrates what Pritchard calls "veritic luck" (Pritchard, *Epistemic Luck*, 146–47) or what we might also call "environmental luck." Environmental luck sometimes seems to undermine knowledge. But it isn't clear that it does here. Do we really want to say that Google searches don't give us knowledge? I don't think so.

11. One might protest that if safety is a requirement for true receptive belief, then since believing X in the scenario is unsafe, then said belief is not receptive. But as I note above about knowledge, this seems anti-intuitive to the extreme. If Web searches—which are paradigm examples of environmental luck—fail to give us receptive beliefs, then we searchers know very much less than we thought we did.

12. Grimm, "Understanding" and "Is Understanding a Species of Knowledge?"

13. This is a broadly Aristotelian account of understanding. See Greco, "Episteme," and Grimm, "Is Understanding a Species of Knowledge?" Not everyone sees understanding as involving knowledge; see Zagzebski, "Recovering Understanding."

14. Thus, understanding need not be factive, although the deeper it becomes, the more it will approach factivity. To understand perfectly, as it were, *is* factive. For further discussion, see Elgin, "Is Understanding Factive?" and Zagzebski, "Recovering Understanding."

15. Ryle, *The Concept of Mind*, 26.
16. Dreyfus and Dreyfus, "A Five-Stage Model" and *Mind over Machine*, especially 30ff.
17. Milner, "Les Troubles de la mémoire."
18. Stanley and Krakauer, "Motor Skill Depends on Knowledge of Facts."
19. For further development of this view, see Stanley, *Know How*.
20. For discussions of this interpretation, see Zagzebski, *On Epistemology*, 141–44.
21. See Plato, *Complete Works*: "Gorgias," X62–63.
22. Dreyfus stresses the importance of experience and motivation for mastery of a skill in *On the Internet*, 42–43.
23. Julie Scelfo, "Kindergarten Shop Class," *New York Times*, March 30, 2011.
24. Boden, *The Creative Mind*, 2–3.
25. Ibid., 2.

Chapter 9: The Internet of Us

1. Rifkin, *The Zero Marginal Cost Society*, 179.
2. Zagzebski, *On Epistemology*, 145. See also "Recovering Understanding."
3. Kitcher, *Abusing Science*. 47–49. I don't mean to suggest that Kitcher would embrace my views on understanding, however.
4. Lazer et al., "The Parable of Google Flu."
5. Solzhenitsyn, *Cancer Ward*, 192.
6. Pete Pachal, "Google Glass Will Have Automatic Picture-Taking Mode," *Mashable*, July 25, 2012. Available at http://mashable.com/2012/07/25/google-glass-photo-mode/#SI4XL.9XkOqI. Accessed September 4, 2015

Bibliography

Achinstein, Peter. *The Nature of Explanation*. Oxford: Oxford University Press, 1983.

Bilton, Nick. *I Live in the Future & Here's How It Works: Why Your World, Work, and Brain Are Being Creatively Disrupted*. New York: Crown, 2010.

Bloom, Paul. "How Do Morals Change?" *Nature* 464, no. 7288 (2010): 490.

———. "The War on Reason." *The Atlantic*, March 2014.

Bloustein, Edward J. "Privacy as an Aspect of Human Dignity: An Answer to Dean Prosser."*New York University Law Review* 39 (1964): 962.

Boden, Margaret A. *The Creative Mind: Myths and Mechanisms*, 2nd edition. New York: Routledge, 2003.

Borges, Jorge Luis. *Ficciones*. Edited by Anthony Kerrigan. New York: Grove Press, 1962.

Bostrom, Nick. "Are We Living in a Computer Simulation?" *Philosophical Quarterly* 53, no. 211 (2003): 243–55.

Brabham, Daren C. *Crowdsourcing*. Cambridge, MA: MIT Press, 2013.

Bibliography

Bruner, Jerome S., and L. E. O. Postman. "On the Perception of Incongruity: A Paradigm." *Journal of Personality* 18, no. 2 (1949): 206–23.

Caldarelli, Guido, and Michele Catanzaro. *Networks: A Very Short Introduction.* Oxford: Oxford University Press, 2012.

Carr, Nicholas. *The Shallows: How the Internet Is Changing the Way We Think, Read and Remember.* New York: Atlantic, 2010.

Cavell, Stanley. *Must We Mean What We Say?* New York: Scribner and Sons, 1969.

Chabris, Christopher F., and Daniel Simons. *The Invisible Gorilla: And Other Ways Our Intuitions Deceive Us.* New York: Broadway Books, 2011.

Clark, Andy. *Natural-Born Cyborgs: Minds, Technologies and the Future of Human Intelligence.* Oxford: Oxford University Press, 2003.

———. *Supersizing the Mind: Embodiment, Action, and Cognitive Extension.* Oxford: Oxford University Press, 2008.

Clark, Andy, and David Chalmers. "The Extended Mind." *Analysis* 58, no. 1 (1998): 7–19.

Coady, David. *What to Believe Now: Applying Epistemology to Contemporary Issues.* New York: John Wiley and Sons, 2012.

Craig, Edward. *Knowledge and the State of Nature.* Oxford: Oxford University Press, 1999.

D'Agata, John, and Jim Fingal. *The Lifespan of a Fact.* New York: W. W. Norton, 2012.

Descartes, René. *Meditations on First Philosophy with Selections from the Objections and Replies.* 1641. Translated by J. Cottingham. Cambridge, UK: Cambridge University Press, 1986.

Dreyfus, Hubert L. *On the Internet,* 2nd edition. New York: Routledge, 2009.

Dreyfus, Hubert L., and Stuart E. Dreyfus. *Mind over Machine.* New York: Simon and Schuster, 2000.

Dreyfus, Stuart E., and Hubert L. Dreyfus. "A five-stage model of the mental activities involved in directed skill acquisition." DTIC Document, 1980.

Elgin, Catherine. "Is Understanding Factive?" In *Epistemic Value*, edited by A. Haddock, A. Miller, and D. Pritchard. Oxford: Oxford University Press, 2009.

Flanagan, Owen J. *Dreaming Souls: Sleep, Dreams, and the Evolution of the Conscious Mind.* New York: Oxford University Press, 2000.

Flanagan, Owen J., and Robert Anthony Williams. "What does the Modularity of Morals Have to Do with Ethics? Four Moral Sprouts Plus or Minus a Few." *Topics in Cognitive Science* 2, no. 3 (2010): 430–53.

Floridi, Luciano. *The Fourth Revolution: How the Infosphere Is Reshaping Human Reality.* Oxford: Oxford University Press, 2014.

Fricker, Miranda. *Epistemic Injustice: Power and the Ethics of Knowing.* Oxford: Oxford University Press, 2007.

Gilbert, Margaret. *Joint Commitment: How We Make the Social World.* Oxford: Oxford University Press, 2015.

———. *Sociality and Responsibility: New Essays in Plural Subject Theory.* Lanham, MD: Rowman and Littlefield, 2000.

Ginsberg, Jeremy, Matthew Mohebbi, Rajan Patel, et al. "Detecting Influenza Epidemics Using Search Engine Query Data." *Nature* 457, no. 7232 (2009): 1012–14.

Gleick, James. *The Information: A History, a Theory, a Flood.* New York: Pantheon Books, 2011.

Goldberg, Sanford C. "The Division of Epistemic Labor." *Episteme* 8, no. 1 (2011): 112–25.

———. *Relying on Others: An Essay in Epistemology.* Oxford: Oxford University Press, 2010.

Goldman, Alvin. *Epistemology and Cognition.* Cambridge, MA: Harvard University Press, 1986.

———. *Knowledge in a Social World.* Oxford: Clarendon Press, 1999.

Bibliography

———. "What Is Justified Belief?" In *Justification and Knowledge*, edited by G. Pappas. Dordrecht: Reidel, 1979.

Gordon, Lewis R. *Disciplinary Decadence: Living Thought in Trying Times*. Boulder, CO: Paradigm, 2006.

———. "Shifting the Geography of Reason in an Age of Disciplinary Decadence." *Transmodernity: Journal of Peripheral Cultural Production of the Luso-Hispanic World*, 1, no. 2 (2011).

Graham, Jesse, Brian A. Nosek, Jonathan Haidt, Ravi Iyer, Spassena Koleva, and Peter H. Ditto. "Mapping the moral domain." *Journal of Personality and Social Psychology* 101, no. 2 (2011): 366–85.

Grau, Carles, Romuald Ginhoux, Alejandro Riera, Thanh Lam Nguyen, Hubert Chauvat, Michael Berg, Julia L. Amengual, Alvaro Pascual-Leone, and Giulio Ruffini. "Conscious Brain-to-Brain Communication in Humans Using Non-Invasive Technologies." *PLoS ONE*, 9, no. 8 (2014): e105225.

Greco, John. "Episteme: Knowledge and Understanding." In *Virtues and their Vices*, edited by Kevin Timpe and Craig A. Boyd. Oxford: Oxford University Press, 2014.

Grimm, Stephen R. "Is Understanding a Species of Knowledge?" *British Journal for the Philosophy of Science* 57, no. 3 (2006): 515–35.

———. "Understanding." In *The Routledge Companion to Epistemology*, edited by Sven Bernecker and Duncan Pritchard, 84-94. London and New York: Routledge.

Haidt, Jonathan. "The Emotional Dog and Its Rational Tail: A Social Intuitionist Approach to Moral Judgment." *Psychological Review*, 108, no. 4 (2001): 814–34.

———. "Reasons Matter (When Intuitions Don't Object)." *New York Times*, October 7, 2012.

———. *The Righteous Mind: Why Good People Are Divided by Politics and Religion*. New York: Pantheon, 2012.

Bibliography

Haslanger, Sally. "Ontology and Social Construction." *Philosophical Topics* 23, no. 2. (1995): 95–125.

———. *Resisting Reality: Social Construction and Social Critique.* Oxford: Oxford University Press, 2012.

Hazlett, Allan. "The Social Value of Non-Deferential Belief." In *Australasian Journal of Philosophy* (2015): DOI: 10.1080/00048 402.2015.1049625.

Hemingway, Mark. "Lies, Damned Lies, and 'Fact-Checking': The Liberal Media's Latest Attempt to Control the Discourse." *Weekly Standard,* December 19, 2011.

Hume, David. *Enquiries Concerning Human Understanding and Concerning the Principles of Morals.* 1777. Edited by L.A. Selby-Bigge and P. Nidditch. Oxford: Oxford University Press, 1975.

Iyengar, Shanto, and Kyu S. Hahn. "Red Media, Blue Media: Evidence of Ideological Selectivity in Media Use." *Journal of Communication* 59, no. 1 (2009): 19–39.

James, William. *Pragmatism and Other Writings.* 1909. New York: Washington Square Press, 1963.

Jeppesen, Lars Bo, and Karim R. Lakhani. "Marginality and Problem-Solving Effectiveness in Broadcast Search." *Organization Science* 21, no. 5 (2010): 1016–33.

Kahneman, Daniel. *Thinking, Fast and Slow.* New York: Farrar, Straus and Giroux, 2011.

Kant, Immanuel. "An Answer to the Question: What Is Enlightenment?" 1784. In *Perpetual Peace and Other Essays*, translated by Ted Humphrey. Indianapolis: Hackett, 1992.

Kitcher, Philip. *Abusing Science: The Case Against Creationism.* Cambridge, MA: MIT Press, 1982.

Kornblith, Hilary. *Knowledge and its Place in Nature.* Oxford: Oxford University Press, 2002.

Kuhn, Thomas S. *The Structure of Scientific Revolutions.* Chicago: University of Chicago Press, 2012.

Kvanvig, Jonathan. *The Value of Knowledge and the Pursuit of Understanding.* Cambridge, UK: Cambridge University Press, 2003.

Lane, Julia, Victoria Stodden, Stefan Bender, and Helen Nissenbaum, eds. *Privacy, Big Data and the Public Good: Frameworks for Engagement.* Cambridge, UK: Cambridge University Press, 2014.

Lazer, David, R. Kennedy, G. King, and A. Vespignani. "The Parable of Google Flu: Traps in Big Data Analysis." *Science* 343 (March 14, 2014): 1203–05.

Levy, Steven. *In the Plex: How Google Thinks, Works and Shapes Our Lives.* New York: Simon and Schuster, 2011.

Locke, John. *An Essay Concerning Human Understanding.* 1690. Edited by P. Nidditch. Oxford: Oxford University Press, 1975.

Lynch, Michael P. *In Praise of Reason: Why Rationality Matters for Democracy.* Cambridge, MA: MIT Press, 2012.

———. *True to Life: Why Truth Matters.* Cambridge, MA: MIT Press, 2004.

Mayer-Schönberger, Viktor, and Kenneth Cukier. *Big Data: A Revolution That Will Transform How We Live, Work, and Think.* Boston: Houghton Mifflin, 2013.

Mercier, Hugo, and Dan Sperber. "Why Do Humans Reason? Arguments for an Argumentative Theory." *Behavioral and Brain Sciences* 34, no. 2 (2011): 57–74.

Milner, Brenda. "Les Troubles de la mémoire accompagnant des lésions hippocampiques bilatérales." In *Physiologie de l'Hippocampe,* edited by P. Passouant, 257–72. Paris: Centre National de la Recherche Scientifique, 1962.

Nietzsche, Friedrich. "On Truth and Lie in an Extra-Moral Sense." In

Bibliography

The Portable Nietzsche, edited and translated by Walter Kaufmann, 42–47. New York: Penguin, 1977.

Nissenbaum, Helen. *Privacy in Context: Technology, Policy, and the Integrity of Social Life.* Palo Alto, CA: Stanford University Press, 2009.

Nozick, Robert. *Anarchy, State, and Utopia.* New York: Basic Books, 1974.

Paxton, Joseph M., Leo Ungar, and Joshua D. Greene. "Reflection and Reasoning in Moral Judgment." *Cognitive Science* 36, no. 1 (2012): 163–77.

People v. Hall, 4 California Supreme Court 399. 1854.

Pew Research Center. "Political Polarization and Media Habits." 2014. Available at: http://www.journalism.org/2014/10/21/political-polarization-media-habits/.

Plato. *Plato: Complete Works.* Edited by John Cooper and D. S. Hutchinson. Indianapolis: Hackett, 1997.

Popper, Karl. *The Open Society and its Enemies.* Vol. 1, *The Spell of Plato.* 1945. London: Routledge and Kegan Paul, 1995.

Priest, Dana, and William Arkin. *Top Secret America: The Rise of the New American Security State.* New York: Little, Brown, 2011.

Pritchard, Duncan. "Anti-Luck Epistemology" *Synthese* 158, no. 3 (2007): 277–97.

———. *Epistemic Luck.* Oxford: Clarendon Press, 2005.

Putnam, Robert D. "Bowling Alone: America's Declining Social Capital." *Journal of Democracy* 6, no. 1 (1995): 65–78.

Rawls, John. *Political Liberalism.* New York: Columbia University Press, 1993.

Rifkin, Jeremy. *The Zero Marginal Cost Society: The Internet of Things, the Collaborative Commons, and the Eclipse of Capitalism.* New York: Palgrave Macmillan, 2014.

Roberts, Paul. *The Impulse Society: America in the Age of Instant Gratification.* New York: Bloomsbury, 2014.

Bibliography

Rose, David. *Enchanted Objects: Design, Human Desire, and the Internet of Things.* New York: Simon and Schuster, 2014.

Rudder, Christian. *Dataclysm.* New York: Crown, 2014.

Russell, Bertrand. "The Expanding Mental Universe." *Saturday Evening Post* 322, no. 3 (1959).

Ryle, Gilbert. *The Concept of Mind.* London: Hutchinson, 1949.

Sajuria, Javier, J. van Heede-Hudson, D. Hudson, Niheer Dasandi, and Y. Theocharis. "Tweeting Alone? An Analysis of Bridging and Bonding Social Capital in Online Networks." *American Politics Research* 43, no. 4 (2014): 1–31.

Scalia, Antonin (dissenting). *Maryland v. Alonzo King, Jr.* 569 United States Supreme Court 1–18. 2013.

Schich, Maximilian, C. Song, Y. Ahn, A Mirsky, M. Martino, A. Barabasi, and D. Helbing. "A Network Framework of Cultural History." *Science* 345, no. 6196 (2014): 558–62.

Schuster, Jack H., and Martin J. Finkelstein. *The American Faculty: The Restructuring of Academic Work and Careers.* Baltimore: Johns Hopkins University Press, 2006.

Shannon, Claude Elwood. "A Mathematical Theory of Communication." *ACM SIGMOBILE Mobile Computing and Communications Review* 5, no. 1 (2001): 3–55.

Silverman, Craig, ed. *Verification Handbook: An Ultimate Guideline to Digital Age Sourcing for Emergency Coverage.* Netherlands: European Journalism Centre, 2014.

Solzehnitsyn, Aleksandr. *Cancer Ward.* 1968. New York: Farrar, Straus and Giroux, 1991.

Sosa, Ernest. "How to Resolve the Pyrrhonian Problematic: A Lesson from Descartes." *Philosophical Studies* 85, no. 2–3 (1997): 229–49.

Bibliography

———. *Reflective Knowledge: Apt Belief and Reflective Knowledge, Volume 2.* Oxford: Oxford University Press, 2011.

———. *A Virtue Epistemology: Apt Belief and Reflective Knowledge, Volume 1.* Oxford: Oxford University Press, 2007.

Stanley, Jason. *How Propaganda Works.* Princeton: Princeton University Press, 2015.

———. *Know How.* Oxford: Oxford University Press, 2011.

Stanley, Jason, and John W. Krakauer. "Motor Skill Depends on Knowledge of Facts." *Frontiers in Human Neuroscience* 7 (2013): 1–11.

Strawson, P. F. *Freedom and Resentment and Other Essays.* New York and London: Routledge, 1998.

Sunstein, Cass R. *Infotopia: How Many Minds Produce Knowledge.* New York: Oxford University Press, 2006.

———. *Republic.com 2.0.* Princeton: Princeton University Press, 2009.

Surowiecki, James. *The Wisdom of Crowds.* New York: Random House, 2005.

Turkle, Sherry. *Alone Together: Why We Expect More from Technology and Less from Each Other.* New York: Basic Books, 2012.

Uhlmann, Eric L., David Pizarro, David Tannenbaum, and Peter Ditto. "The Motivated Use of Moral Principles." *Judgment and Decision Making* 4, no. 6 (2009): 476–91.

Weinberger, D. *Too Big to Know: Rethinking Knowledge Now That the Facts Aren't the Facts, Experts Are Everywhere, and the Smartest Person in the Room Is the Room.* New York: Basic Books, 2011.

Wieseltier, Leon. "Among the Disrupted." *New York Times Sunday Book Review*, January 7, 2015.

Wittgenstein, Ludwig. *Philosphical Investigations.* Translated by G. E. M. Anscombe. Oxford: Basil Blackwell, 1963.

Bibliography

Wright, Crispin. "Scepticism and Dreaming: Imploding the Demon." *Mind*, 100, no. 1 (1991): 87–116.

Yardi, Sarita, and Danah Boyd. "Dynamic Debates: An Analysis of Group Polarization over Time on Twitter." *Bulletin of Science, Technology and Society* 30, no. 5 (2010): 316–27.

Zagzebski, Linda. *On Epistemology*. Belmont, CA: Wadsworth, 2008.

———. "Recovering Understanding." In *Knowledge, Truth and Duty: Essays on Epistemic Justification, Responsibility and Virtue*, edited by M. Steup. Oxford: Oxford University Press, 2001.

———. "What is Knowledge?" In *The Blackwell Guide to Epistemology*, edited by J. Greco and E. Sosa, 92–116. Oxford: Blackwell, 1999.

Index

Page numbers in *italics* refer to illustrations. Page numbers beginning with 191 refer to endnotes.

Index

Index

Index

Index

Index

Index

Index

Index

Index

Index

Index

Index

Index

in problem solving, 136
use of term, 111–12
neural system, 26
neural transplants, 3, 5
Neurath, Otto, 128–29
neuromedia, 3–5, 12, 17–19,
  113–14, 132, 149, 168, 180–
  82, 184
  limitations of, 174
  as threat to education, 153–54
Newton, Isaac, 175
New Yorker, 25, 26
New York Times, 122, 174
Nietzsche, Friedrich, 111
Nobel laureates, 149
noble lie, 83, 86
nonfiction, 79–80
NPR, 78, 80
NSA:
  alleged privacy abuses by,
    98–100, 138
  data mining by, 9, 91, 95–96, 108,
    167
  proposed limitations on, 109
Ntrepid, 81
nuclear weapons technology, xvii
nullius in verba (take nobody's
  word for it), 34

Obama, Barack, 7, 100
  administration, 109

objectivity, objective truth, 45, 74
  as anchor for belief, 131
  in constructed world, 83–86
  as foundation for knowledge, 127
observation, 49, 60
  affected by expectations, 159–
    60
  behavior affected by, 91, 97
"oceanic feeling," 184
"offlife," 70
OkCupid, 157
"onlife," 70
online identity creation, 73–74
online ranking, 119–21, 136
open access research sharing sites,
  135–36
open society:
  closed politics vs., 144–45
  values of, 41–43, 62
open source software, 135
Operation Earnest Voice, 81
Operation Ivy, ix
opinion:
  knowledge vs., 13, 14, 126
  in online ranking, 119–20
  persuasion and, 50–51
  truth as constructed by, 85–86
optical illusions, 67
Oracle of Delphi, 16–17, 171
Outcome-Based Education (OBE),
  61–62

Index

Index

Index

Index

Index

Index

About the Author

Michael Patrick Lynch is a professor of philosophy and the director of the Humanities Institute at the University of Connecticut. He is the author or editor of more than seven books, including the *New York Times Book Review* Editors' Choice *True to Life*, as well as *Truth as One and Many*. The recipient of the Medal for Research Excellence from the University of Connecticut's College of Liberal Arts and Sciences, Lynch has held grants from the National Endowment for the Humanities and the Bogliasco Foundation among others. He is a frequent contributor to the *New York Times'* The Stone series. An avid sailor, he lives in rural Connecticut.